Aids to Microbiology and Im

Aids to Microbiology and Infectious Diseases

R. A. V. Benn

FRACP   FRCPA Dip Bact (Lond)

Microbiologist
Royal Prince Alfred Hospital, Sydney, Australia

CHURCHILL LIVINGSTONE
EDINBURGH LONDON MELBOURNE AND NEW YORK 1985

CHURCHILL LIVINGSTONE
Medical Division of Longman Group Limited

Distributed in the United States of America by
Churchill Livingstone Inc., 1560 Broadway,
New York, N.Y. 10036, and by associated
companies, branches and representatives
throughout the world.

First published/edition 1985

ISBN 0 443 03127 4

British Library Cataloguing in Publication Data
Benn, R.A.V.
    Aids to microbiology and infectious diseases.
    1. Communicable diseases
    I. Title
    616.9   RC111

Produced by Longman Singapore Publishers (Pte) Ltd.
Printed in Singapore.

Preface

For students who like to prepare for examinations using lists and tables, this book adds another medical specialty to a popular series. As taught in most medical schools, microbiology is a pre-clinical science and infectious diseases a sub-specialty of internal medicine. In attempting to associate the two, as many of us think should be the case, this outline blends the facts of both disciplines in most chapters.

The book is intended for undergraduates and trainee physicians as well as those entering specialties such as surgery and obstetrics in which the management of infection is an important component. There is no alternative to reading the whole book, but students of clinical medicine will clearly not need to concern themselves with the laboratory and taxonimic matters listed in some of the tables in the earlier chapters.

Sydney, 1985   R.A.V.B.

Contents

Classification, diagnostic methods

The agents of infectious disease are classified according to size, structure and method of reproduction:

A. METAZOANS (worms)

— Multicellular organisms, complex life cycles, multiple hosts
— Tissue invasion induces eosinophilia
— Classified initially according to external appearance, viz:
  1. Tapeworms (platyhelminths)
  2. Roundworms (nematodes)
  3. Flukes (trematodes)

B. PROTOZOANS

— Single-celled eukaryotes (i.e. DNA is enclosed within a nuclear membrane)
— Classified according to methods of locomotion and reproduction, viz:
  1. Flagellates: *Trichomonas, Giardia, Leishmania, Trypanosoma, Naegleria*
  2. Amoebae: *Entamoeba, Acanthamoeba*
  3. Ciliates: *Balantidium*
  4. Sporozoa: (have cycles of sexual and asexual reproduction): *Plasmodium, Toxoplasma, Cryptosporidium, Isospora*
  5. Unclassified: *Pneumocystis*

(Metazoans and protozoans are collectively known as *parasites*)

C. FUNGI

— Eukaryotes
— Rigid chitinous cell walls
— Fungi of medical interest are best classified morphologically:
  1. Moulds: filamentous, spore-forming
  2. Yeasts: unicellular, reproduce by budding
  3. Dimorphic: grow as moulds or yeasts, depending upon the cultural conditions.

D. BACTERIA

— Unicellular, prokaryotic (i.e., have no nuclear membrane), reproduce by binary fission, 0.2-1.5 um in short diameter, lack mitochondria.
— The rigid cell wall contains peptidoglycan.
— Classified according to size, shape, cell wall structure, habitat and aerotolerance:
1. 'Higher bacteria': Complex cell walls, some are filamentous and branched—*Mycobacteria, Nocardia, Actinomyces, Streptomyces* and *Corynebacteria*
2. 'eubacteria': commonly encountered aerobic and facultative anaerobic bacteria.
3. Strict anaerobes: grow only in low oxygen concentration
4. Spirochaetes: helically coiled around an axial filament. Slender and flexuous
5. Mycoplasmas: lack rigid cell walls
6. Chlamydiae; rickettsiae: multiply only in host cells

E. VIRUSES

— Particles of DNA or RNA enclosed within an integrated shell of protein, of defined symmetry
— Reproduce by replication inside living cells using the host's synthetic machinery
— Initial classification is made according to type of nucleic acid:
1. DNA viruses: pox; herpes; adeno; hepadna; papova- and parvoviruses
2. RNA viruses: orthomyxo; paramyxo; rhabdo; reo; toga; picorna; filo- and arenaviruses.

TAXONOMY

The science of classification and nomenclature of living things (Greek: *tasso* = to arrange; *nomia* = distribution).

Creates groups such as (in diminishing size) orders, families, genera and species.

The interest of the medical microbiologist in taxonomy lies in the identification of microbial isolates which are pathogenic (i.e. cause disease).

In the Linnaean system, an organism is named according to its genus and species (e.g. *Staphylococcus aureus*).

Definition of a species

Amongst eukaryotes (metazoans, protozoans, fungi), a species is defined as a group of organisms which is able to reproduce sexually.

In prokaryotes there is no sexual union so that a species must be defined in other ways, viz:

A. A group of organisms which bear a close resemblance as determined by a large range (about 50–200) of biochemical and other tests. These tests can be much influenced by phenotypic variation, mutation and presence of plasmids.
B. A group with very similar DNA composition as determined by genome size, proportion of guanine plus cytosine in total nucleic acids, percentage of DNA homology in hybridisation tests.

A genus (plural, genera)
— is a group of phenotypically similar species
— DNA relatedness is usually in the order of 40–60%

Types, subspecies and varieties
— are groups within a single species
— are based on methods such as the following:
  1. *serotypes*: cell surface/wall antigen
  2. *'phage types'*: bacteriophage susceptibility
  3. *biotypes*: biochemical tests
  4. *auxotypes*: growth requirements
  5. *restriction enzyme types*: electrophoretic patterns of DNA fragments following hydrolysis

DIAGNOSTIC METHODS IN MICROBIOLOGY

1. Microscopy

a. *light microscopy*
  — is used in conjunction with stains of the bacterial cell
  — can only magnify about 1000 times
  — detects organisms of diameter 0.2 μm or greater

b. *phase contrast microscopy*
  — converts difference in phase into differences in light intensity
  — differentiates internal structures of living cells

c. *dark field microscopy*
  — illuminates the specimen in such a way that only light reflected from the specimen is seen by the observer
  — detects slender bacteria with characteristic motility (e.g. spirochaetes)

d. *electron microscopy*
  — uses a beam of electrons focused by magnets
  — can resolve particles with a diameter of 0.001 μm

e. *immunofluorescence microscopy*
  — using ultraviolet light, detects fluorescent compounds conjugated to antibodies

— is performed by flooding a fixed smear of the specimen with fluorescent antibodies in the search for bacterial or viral antigens

Staining methods for light microscopy
a. Gram: differentiates two types of bacterial cell wall
b. Giemsa: detects most bacteria, differentiates human cell types
c. Silver: deposited out of solution on bacteria and their appendages. Demonstrates flagella. Useful for organisms which do not stain well with Gram method (e.g., fungi, *Legionella*, cysts of *Pneumocystis carinii*)
d. Methelene blue: stains most bacteria, detects metachromatic granules
e. Carbol-fuchsin (Ziehl-Neelsen stain): at 1%, together with heating, stains mycobacteria and bacterial spores.
f. Iodine: used as a mordant in Gram's method. Stains eggs in faeces

2. Detection of structural components or products

a. Antigens
Examples of the use of antigen detection are shown in Table 1.

Table 1

Agent	Site of detection	Serological methods
Hepatitis B	Serum	Radioimmunoassay Enzyme-linked immunoassay
Cryptoccus neoformans	CSF, serum	Latex agglutination
Pneumococcus Haemophilus Meningococcus	CSF	Immunoelectrophoresis Coagglutination
Legionella	Sputum	Immunofluorescence

b. Fermentation products
— especially of anaerobic bacteria, may be distinctive
— detected by chromatography

c. Toxins
— e.g. use of mice to detect botulinus toxin in serum

d. Detection of nucleic acids by hybridisation
— has been recently introduced in research laboratories for detection of viral and plasmid nucleic acids

3. Propogation (i.e. cultivation) of the organism
a. Inert culture medium — for most bacteria and fungi
b. Cell culture — for viruses and chlamydia
c. Embryonated eggs — for some viruses and rickettsiae
d. Animals — for some viruses and rickettsiae

4. Detection of antibody response by host following tissue invasion
— takes several days to appear
— involves numerous techniques of varying sensitivity, specificity and complexity:

a. *agglutination*
— whole bacterial cells are used as antigen to clump in the presence of antibodies
— e.g. Widal (typhoid), Weil–Felix (rickettsia)

b. *flocculation*
— of insoluble macromolecules (e.g. VDRL test for syphilis)

c. *precipitation*
— of soluble antigen when it forms a complex with antibody, in solution or in agar gel

d. *immunoelectrophoresis*
— enhances precipitation in agar gel by forcing the migration of antigen and antibody towards each other in an electric field

e. *passive agglutination*
— uses a carrier particle for the antigen:
  (i) red cells: haemagglutination
  (ii) latex: latex agglutination

f. *complement fixation (CFT)*
— antigen-antibody formation fixes and depletes complement included in the test system
— absence of free complement is detected by then adding rabbit cells and their haemolytic antibodies

g. *immunofluorescence (IFA)*
— detects antibodies attached to antigens of cells fixed to glass slides

h. *radioimmunoassay (RIA)*
— uses radioactive tracer to detect antibodies attached to antigens fixed to solid surfaces of test tubes, latex beads, etc

i. *enzyme-linked assay (ELISA)*
— as radioimmunoassay using enzyme-linked tracer

j. *neutralisation*
— of the growth of viruses or mycoplasmas in the presence of their antibody

k. *haemagglutination-inhibition (HAI)*
— prevention of the capacity of certain viruses to agglutinate red cells in the presence of their antibody

l. *immobilisation*
— of motile organisms in the presence of their antibodies
— e.g. *Treponema pallidum* immobilisation test

m. *animal protection*
— ability of antibodies to prevent the lethal effects of toxins or viruses

n. *inhibition of enzymes produced by bacteria*
— e.g. streptolysin O, streptococcal DNAse

Virulence and host defences

VIRULENCE (pathogenicity)

The ability of microbes to produce disease
Mechanisms involved include:
1. Attachment to mucosal epithelium or skin
2. Invasion and multiplication
3. Resistance to phagocytosis or intracellular killing
    — dependes upon bacterial surface proteins or polysaccharides
4. Localisation in certain tissues
5. Toxin production
    a. *Exotoxins*—proteins which diffuse into the media
    surrounding bacteria. Include the following:
        — Cytolysins: destroy cells
        — Neurotoxins: interfere with neurotransmission at synapses
        — Enterotoxins: activate adenycyclase of small bowel
        mucosal cells
    b. *Endotoxins*—surface components of gram negative cell walls.
    Activate complement. Cause release of endogenous pyrogen

A comparison of endotoxins and exotoxins is shown in Table 2.

HOST DEFENCES

1. Anatomical barriers

a. Skin
— Fatty acids, low pH

b. Respiratory tract
— Ciliated epithelium
— Coughing
— Secretory IgA
— Lysozyme: cleaves bacterial glycopeptide

c. Gut
— Peristalsis
— Low pH (stomach)
— Secretory IgA

Table 2 A comparison of two types of toxin

	Endotoxins	Exotoxins
Bacteria	Gram negative	Gram positive
Composition	Phospholipid Polysaccharide	Protein
Origin	Cell wall	Cytoplasm
Heat stability	Yes	No
Mode of action	Complement activation, kinin release	Specific cytotoxins
Nature of toxin	Lipid	Protein
Antigenic component	Polysaccharide	Protein
Toxoid formation	No	Yes
Toxicity	Weak	Strong

d. Female genital tract
— Low pH (during child-bearing years)
— Secretory IgA

2. **Normal surface flora**
— competes with pathogens for nutrients and receptors on host cells
— produces antimicrobial agents (bacteriocins, antibiotics)
— stimulates cross-reactive antibodies to pathogens

3. **Humoral immunity** (antibodies and complement)

4. **Phagocytosis**
— Engulfment of microbes by granulocytes and macrophages.
— Initiated by presence of bacterial products, antigens, immune mechanisms and complement activation.
— Sequence of events is as follows:
   a. phagocytes adhere to vascular epithelium
   b. diapedesis (migration through vascular epithelium)
   c. chemotaxis (movement through extracelluar space to site of initiating process)
   d. attachment to microbe
   e. ingestion into phagocyte
   f. degranulation:
      — discharge of the contents of the granule into the phagosome

- accompanied by a burst of metabolic activity, oxygen consumption, superoxide and hydrogen peroxide formation, stimulation of HMP shunt

g. bacterial destruction

5. Cell-mediated immunity

- Sensitisation of lymphocytes by antigens
- Transformation in the presence of antigen results in:
  a. secretion of soluble substances (lymphokines) which influence macrophage and neutrophil activity. Activated macrophages release monokines
  b. capability of certain lymphocytes to destroy cells containing organisms, especially viruses (killer cells)
  c. induction of B cell response (helper cells) — IgM receptors
  d. suppression of immunologic phenomena (suppressor cells) — receptor for Fc portion of IgG

6. Interferons

- heterogenous glycoproteins produced in various cell types in response to the presence of viruses, double-stranded RNA, bacterial endotoxin etc
- bind to cell membranes and stimulate production of proteins which inhibit translation of viral messenger RNA

Epidemiology, methods of transmission of infectious disease

METHODS OF TRANSMISSION OF MICROORGANISMS

1. **Autogenous**
   - occurs when the patient is the source of the organism causing the infection
   - is the origin of infections such as the following:
   - Surgical wound infections: staphylococci; bowel organisms
   - Urinary tract infections: bowel coliforms
   - Postnatal endometritis: vaginal flora, especially anaerobes
   - Bronchopneumonia: *Pneumococcus*; *Haemophilus influenzae*

2. **Contact**
   - by hands, instruments, clothes, contaminated solutions

3. **Ingestion**
   - of intestinal pathogens (faeco-oral spread)
   - usually in food or water, following multiplication of the pathogen in the vehicle of transmission

4. **Air**
   - acts as a vehicle for transmission especially of organisms which are naturally dispersed in air and/or which resist drying, e.g.:
   Gram-positive bacteria: staphylococci, streptococci, *Corynebacterium diphtheriae*, mycobacteria, *Bacillus anthracis*
   Respiratory viruses: agents causing common colds
   Certain gram negative bacilli: legionellae, *Bordetella pertussis*, *Yersinia pestis*
   Viruses causing childhood exanthems: varicella, measles, rubella, mumps
   - Microoorganisms may be transported in air by:
     a. Droplets expelled from the respiratory tract by sneezing or coughing. Droplet nuclei are the residue left after evaporation of droplets
     b. Dust particles
     c. Skin scales

— act as a method of dispersal of staphylococci which are disseminated from carriers on skin scales, 12–15 μm in diameter
d. Aerosols created by nebulisers, sprays, cleaning equipment, etc

5. **Venereal**
— following intimate mucosal contact, viz. genital, oral-genital, anal-genital, oral-anal (male homosexuals)

6. **Parenteral**
— injections, blood transfusions

7. **Insects** (arthropod-borne)

8. **Vertical** (congenital)
— maternal infection of the fetus via the placenta (as opposed to 'horizontal' infection from person to person)

METHODS FOR THE PREVENTION OF TRANSMISSION OF INFECTION

1. *Autogenous*
   Pre-operative skin antiseptics, antibiotics
2. Contact
   Handwashing
   Gloves
   Sterilisation of instruments, fomites
3. Ingestion
   Heating and refrigeration of food
   Filtration and chlorination of drinking water
   Inspection and testing of food
   Handwashing
4. Air
   Air-conditioning and filtration
   Facemasks
   Avoidance of crowding
   Patient isolation
5. Venereal
   Condoms
6. Parenteral
   Sterilisation of intravenous solutions and equipment
   Screening of blood donors
7. Insects
   Insect repellant
   Elimination of breeding sites
   Malarial prophylaxis
8. Vertical
   Screening of pregnant women

ZOONOSES

— are infections naturally transmitted between vertebrate animals and man
— may be spread by some of the methods listed above (contact, air, ingestion, arthropods)
— include a large number of infections from all microbial classes usually acquired in a rural environment

Infections transmitted by milk

1. Tuberculosis
2. Brucellosis
3. Q fever
4. Yersiniosis
5. Salmonellosis
6. Campylobacter

Infections transmitted by water (Table 3)

Table 3

	Disease	Organisms
Enteric pathogens	Gastroenteritis	Vibrio *Salmonella* *Giardia* *Shigella* *Campylobacter* Rotavirus Parvovirus
Environmental organisms	Dermatitis	*Pseudomonas aeruginosa* *Schistosomiasis*
	Meningitis	*Naegleria fowleri*
	Pneumonia	*Legionella* *Pseud. pseudomallei*
	Wound infection	Halophilic vibrios

Bacterial classification, structure and diagnostic methods

CLASSIFICATION

Bacteria are classified initially according to:

1. Shape
a. Cocci: spheres
b. Bacilli: rods
c. Vibrios: curved rods
d. Spirochaetes: coils

2. Gram reaction
— depends upon the ability of some bacteria (designated gram positive) to retain basic dyes, such as gentian violet, despite washing with lipid solvents such as alcohol or ether
— reflects differing cell wall composition (see below)

3. Aerotolerance
Ability of bacteria to grow in various concentrations of oxygen, as follows:
a. Strict aerobes: grow only in the presence of oxygen
b. Facultative aerobes: grow with or without oxygen
c. Microaerophilic bacteria: grow only in reduced oxygen concentration, often requiring CO_2, as well.
d. Strict anaerobes: grow only in very low concentrations of oxygen.

4. Spores
— are formed within the bacterial cell
— resist heat, drying and antiseptics
— are present in only two genera of medical interest:
   a. *Bacillus* species — aerobic bacilli
   b. *Clostridium* species — anaerobic bacilli

5. Motility
Bacteria move with the aid of flagella, one or more long hair-like projections containing contractile protein.
Detected by:
a. Hanging drop preparations examined by light, phase or dark ground microscopy
b. Stains revealing flagella, e.g. silver
c. Electron microscopy
d. Movement of growing culture in semisolid agar
The disposition of flagella is used in classification as follows:
a. Polar: originate from one or both ends of the cell
b. Peritrichous: originate from the sides of the cell

6. Production of acid from glucose
The metabolic activity of the cell is designated:
a. 'fermentative': acid is produced under anaerobic conditions
b. 'oxidative': acid is produced only in aerobic culture
(the Hugh and Leifson test uses 0.2% glucose, a weak buffer and paraffin overlay to detect this difference)
   Fermentative organisms produce acid from 1% glucose on solid or in liquid media. Suitable indicators include bromthymol blue, phenol red and neutral red.
   The nature of fermentation products is used in the classification of some anaerobes.

7. Acid-fastness
The presence of mycolic acids in the cell walls of mycobacteria and nocardias enables these cells to retain dyes despite washing with acid.

8. Production of certain enzymes

a. Oxidase
— is detected using 1% tetramethyl-p-phenylenediamine solution which is reduced by bacteria containing electron- transporting oxidases to a deep purple colour.

b. Catalase
— is detected as the production of oxygen from 10 vol hydrogen peroxide solution

c. Reduction of nitrate to nitrite and nitrogen (anaerobic respiration)

BACTERIAL STRUCTURE

A. Cell envelope

1. Cell membrane
— consists of phospholipid
— contains enzymes for active transport as well as electron transport (bacteria do not have mitochondria)

2. Cell wall — has several constituents:
a. peptidoglycan (murein)
   — is a unique structural compound of bacteria, consisting of:
   (i) muramic acid plus N-acetylglucosamine linked by B-1, 4 bonds
   (ii) side chains of alternating D and L pentapeptides
b. teichoic acid
   — occurs predominantly in Gram-positive bacteria
c. lipopolysaccharide ('endotoxin')
   — occurs only in Gram-negative bacteria

3. Capsule
— varies in composition and thickness
— protects the cell from the ingestion by phagocytes

B. Appendages

1. Flagella
— are long filamentous projections about 0.02 µm in diameter originating in the cell cytoplasm
— contain contractile proteins (flagellin) which propel the organism

2. Pili (fimbriae)
— are hairlike projections of 4–8 nm diameter present around the cell of numerous gram negative bacilli
— promote adhesion to solid surfaces including red cells and other bacteria

C. Cytoplasmic structures

1. Ribosomes

2. Mesosomes

3. Granular inclusions

D. Nuclear material (bacteria have no nuclear membrane)

1. Chromatin
— consists of coiled circular haploid DNA attached to the cell membrane

2. Plasmids
— extrachromosomal DNA

E. Spores

Table 4 The *Gram stain* distinguishes two classes of bacteria.

	Gram-positive	Gram-negative
Wall thickness	150–800 nm	100 nm
Peptidoglycan	+++	+
Teichoic acid	++	±
Lipopolysaccharide	–	++
Relatively resistant to	Heat Drying Lipid solvents	Dyes Detergents

BACTERIAL CULTURE

Using methods which have changed little since the turn of the last century, bacteria are cultured:
1. In liquid media
    — to detect small numbers of organisms in body fluids (e.g. blood)
    — to encourage, in mixed culture and in the presence of selective inhibitors, the growth of one organism at the expense of others (enrichment), e.g. *Salmonella* in faeces
    — to dilute out the effect of antagonists, such as antibiotics and disinfectants
    — to perform metabolic tests (e.g. carbohydrate fermentation)
2. On solid media
    — to select colonies derived from single cells.
    — to observe colonial appearance, haemolysis of blood etc

Components of bacterial culture media
1. Water, electrolytes, buffers, minerals — to approximate the composition of extracellular fluid at pH 7.2–7.4
2. Protein hydrolysate made from casein, meat, brain or soya bean
3. Solidifying agent
    a. agar
        — is a polysaccharide derived from seaweed, not decomposed by most human bacteria

— forms a gel which melts at 95°C and solidifies at 45°C
Thermolabile compounds can be added at 50°C, just above that
resulting in gel formation.
  b. egg yolk, solidified by heating (e.g. Lowenstein–Jensen
     medium for mycobacteria)
  c. serum, solidified by heating (e.g. Loeffler's medium for
     Corynebacterium diphtheriae)
4. Additives to provide growth factors for fastidious organisms
  a. blood (defibrinated)
     — is added at a concentration of 5% (sheep, horse, human,
       rabbit) to media routinely used in medical laboratories
     — heating releases haeme and NAD (chocolate blood agar)
  b. haemin, lysed blood, Filde's digest of blood
     — may be used instead of whole blood if a clear medium is
       required
  c. yeast extract
     — provides supplementary amino acids, carbohydrates and
       B vitamins.
  d. compounds to remove toxic fatty acids
     — e.g. charcoal, starch
  e. carbohydrates
     — promote growth of organisms such as fungi
  f. selective agents
     — inhibit the growth of some bacteria at the expense of
       others (Table 5)

Table 5

Selective agent	Specimen	Selects out	Inhibits
Bile	Urine	Coliforms	Vaginal flora
Crystal violet	Throat	Coliforms Streptococci	Staphylococci
7% NaCl	Nose, throat	Staphylococci	Coliforms
Cefoxitin	Faeces	*C. difficile*	Coliforms, other anaerobes
Cetrimide	Faeces	Pseudomonads	Coliforms
Cephalexin	Throat	*B. pertussis*	Oral flora

BACTERIAL GENETICS

The organisation of genetic elements within the bacterial cell is as
follows:

1. Chromosome
— is singular, circular, haploid and held in helical supercoils (whose structure is created by DNA gyrase) by RNA and protein
— is not separated from the rest of the cell by a nuclear membrane
— is replicated continuously during rapid cell division and segregated by attachment to the growing cell wall
— does not undergo sexual (i.e. meiotic) recombination with chromosomes of other cells

2. Extrachromosomal DNA

a. Plasmids
— are small circular pieces of naked DNA present in the cell cytoplasm, constituting about 1–2% of total cell genetic material
— are not essential for growth but may code for such phenomena as antibiotic resistance, sex factors, toxins prodution, bacteriocins, and fermentation of carbohydrates
— multiply independently of host DNA
— may unite with chromosome, in which case there is a high frequency of transfer to other cells
— can be removed by dyes
— may be transferred to other cells, including unrelated species, along sex pili (conjugation)
— have proven to be ideal vectors for molecular cloning of foreign genes

b. Bacteriophages
— are bacterial viruses
— may multiply and destroy the cell or integrate with the bacterial chromosome (lysogeny)

3. Transposons
— are pieces of DNA capable of transfer from one replicon to another
— do not require the rec A recombination system
— may produce mutations and DNA rearrangements
— may be randomly or preferentially inserted into recipient replicon
— have inverted repeat sequences at both ends
— create diversity amongst plasmids, join together DNA segments with little ancestral relationship

MECHANISMS OF GENETIC CHANGE IN BACTERIA

1. Mutation
— results from the change in nucleotide sequence within a chromosomal gene
— occurs spontaneously once every 10^5–10^{10} cell divisions but rate may be inreased by exposure to mutagens such as ultraviolet irradiation and chemicals (as used in the Ames test)

2. Recombination (of nucleic acid)

a. *Transformation*
— involves transfer of free DNA from chromosome to chromosome.
— the base sequences of the absorbed DNA must be compatible with that of the host

b. *Transduction*
— requires lysogenic bacteriophage to transfer a piece of the host's chromosome to that of the recipient
— demonstrated in relatively few species, especially staphylococci

c. *Conjugation*
— transfers plasmids from host to recipient, which need not be the same species

Table 6 Summary of methods of genetic recombination

	Method of transfer	*No. of genes transferred*	*Frequency of occurrence*	*Relatedness of transfer*
Transformation	Free DNA	2–10	? +/−	Same
Transduction	Lysogenic bacteriophage	10–25	+	Same
Conjugation	Sex pili	25–100	+++	Quite variable

Gram-positive cocci

Three genera are of medical interest (Table 7)

Table 7

	Staphylococcus	*Micrococcus*	*Streptococcus*
Morphology	Clusters	Tetrads	Chains
Metabolism	Fermentative	Oxidative	Fermentative
Sensitivity to dyes	Sensitive	Sensitive	Resistant
Catalase production	+	+	− (Contain no haeme compounds)
Nutritional requirements	Complex	Variable	Complex
Aerotolerance	Facultative	Strict aerobes	Facultative
Habitat	Skin	Air, skin	Skin, mucous membranes
Growth in 7% NaCl	Yes	Yes	No
Lysis by lysostaphin	Yes	No	No
Lysis by lysozyme	No	Yes	No

GENUS STAPHYLOCOCCUS

Production of coagulase distinguishes *Staph aureus* from less pathogenic members of the genus. Methods of detection are as follows

1. **Bound coagulase** (clumping factor)
   — present on the cell surface of 80–90% of strains of *Staph aureus*
   — causes organisms to clump in the presence of fibrinogen (in plasma)
   — used as a rapid slide test

2. **Free coagulase**
   — activates prothrombin in plasma
   — performed as an overnight tube test.

Table 8 Antigens of *Staph aureus*

Surface	Capsule	
	Protein A	Binds Fc portion of IgG
Cell wall	Peptidoglycan	
	Ribitol teichoic acid	
Extracellular substances	Haemolysins	Alpha, beta and gamma
	Coagulases	(see above)
	Leucocidins	
	Lipases	
	DNA-ase	
	Phosphatases	
	Proteases	
	Hyaluronidases	
Toxins	Dermatotoxin	
	Enterotoxins	

Spectrum of disease due to *Staph aureus*

1. *Invasive*
   — Furunculosis (boils, carbuncles, styes)
   — Wound infection (traumatic and surgical)
   — Endocarditis
   — Pneumonia
   — Abscess formation in deep tissues
     Osteomyelitis
     Septic arthritis
     Perinephric abscess (renal carbuncle)
     Brain abscess
     Suppurative myositis (in the tropics)
   — Septicaemia

2. *Toxic*
   — Ritter's disease   ⎫
   — Lyell's syndrome ⎬ toxic epidermal necrolysis
   — Food poisoning   ⎭
   — Toxic shock syndrome

Carriage rate of Staph aureus in various populations
1. Nasal vestibule
   Neonates   60–80%
   One year   10–20
   Over 10 year   30–50
2. Perineum
   Adult males   5–10
   Adult females 0–5
3. Colonisation rate rises in
   a. patients with skin diseases
      — dermatitis, psoriasis, wounds
   b. patients receiving frequent injections:
      — diabetics, drug addicts, haemodialysis
   c. patients hospitalised for long periods

Table 9 Medically important staphylococci

	Staph aureus	Staph epidermidis	Staph saprophyticus
Coagulase	+	–	–
Fermentation of mannitol	+	Variable	Variable
Heat-stable DNA-ase	+	–	–
Protein A in cell wall	+	–	–
Sensitivity to novobiocin	S	S	R
Production of phosphatae	+	Variable	–
Clinical associations	Invasive	Colonisation of surgical prostheses	Cystitis in young women

GENUS STREPTOCOCCUS

Several species are of medical interest. Classified initially using the following tests:
1. *Haemolysis* on 5% blood agar
   Beta haemolysis: complete clearing of red cells around colonies, enhanced by anaerobic incubation, inhibited by dextrose
   Alpha haemolysis: conversion of haemoglobin to green compounds in unlysed red cells around the colonies
   Non-haemolytic: Sometimes incorrectly called 'gamma' haemolysis.

2. *Antigens* of the cell wall (Lancefield group) Group-specific carbohydrate, detected by:
   a. extraction with hot acid, formamide, enzymes or nitrous acid followed by a precipitation test, or
   b. agglutination of whole cells in the presence of group-specific antibodies attached to staphylococci (coagglutination) or latex particles.
3. *Strep. pneumoniae* is identified by its susceptibility to
   — deoxycholate (lyses cells by activating autolytic enzymes)
   — optochin (ethylhydrocuprein) in a disc test

4. *Strep. pyogenes* is susceptible to bacitracin (0.01 unit disc)

Table 10 Classification of streptococci

	Haemolysis	Group antigen	Habitat	Pathogenicity
Strep. pyogenes	Beta	A	Pharynx	Pharyngitis Impetigo Cellulitis Nephritis Rheumatic fever
Strep. agalactiae	Partial beta	B	Bowel Vagina	Neonatal, pneumonia &, meningitis
Streptococcus, group C	Beta	C	Pharynx	Occasional
Streptococcus, group G	Beta	G	Pharynx	pharyngitis & cellulitis
Strep. faecalis	Variable	D	Bowel	Urinary tract infection endocarditis
Strep. milleri	Variable	F, and other	Pharynx	Abscesses, especially liver and brain
Strep. 'viridans'	Alpha	Various	Pharynx	Endocarditis
Strep. pneumoniae	Alpha	A*	Pharynx	Pneumonia Meningitis Otitis
Strep. suis	Variable	R	Pigs	Endocarditis Meningitis (rare)

* The group antigen of *Strep. pneumoniae* is not used to identify the organism

Table 11 Antigens and structures of *Streptococcus pyogenes*

Capsule	Hyaluronic acid	Not antigenic Gives colonies a smooth shiny appearance
Cell wall	M protein	Roughened fimbriae, necessary for virulence. 60 antigenic types, some associated with nephritis.
	T antigen	
	Carbohydrate	Group-specific (group A) polymer
	Inner layer of peptidoglycan	
Extracellular products	Haemolysins	Streptolysin S (oxygen stable)
		Streptolysin O (oxygen labile, makes larger zones of haemolysis when growing anaerobically)
	Streptokinase	
	Nucleases	DNAse, RNAse, DPNase
	Hyaluronidase Proteinases Amylase	
	Erythrogenic toxin	Causes erythema of scarlet fever

Antibodies to extracellular products are useful markers of recent infection (eg, antistreptolysin O, antiDNAse).

Spectrum of diseases caused by *Strep. pyogenes*
1. *Invasive*
   Pharyngitis
   Cellulitis, erysipelas, lymphangitis
   Impetigo
   Wound infection (traumatic and surgical)
2. *Toxic*
   Scarlet fever
3. *Post-streptococcal*
   Rheumatic fever
   Glomerulonephritis
   A comparison of the two diseases is given in Table 12.

Table 12

	Rheumatic fever	Glomerulonephritis
M serotypes	Any	4, 12, 25 and 49 predominate
Pathogenesis	Formation of antibodies cross reactive between streptococci and cardiac tissue	Immune complex deposition
Pathology	Pericarditis Myocarditis Valvulitis Polyarthritis Chorea	Diffuse endocapillary proliferative glomerulonephritis
Recurrences	Frequent	Rare
Prognosis	Recurrence leads to valve damage	Occasional progressive disease
Prevention	Prolonged penicillin administation	Unnecessary

Table 13 Structural components of various streptococci

	Capsule	Surface proteins	Cell wall group substance
Strep. pyogenes	Hyaluronic acid	M protein 60 types	carbohydrate
Strep. agalactiae	Polysaccharide 4 types	Nil	carbohydrate
Strep. faecalis	Nil	Nil	teichoic acid
Strep. pneumoniae	Polysaccharide 86 types	Nil	carbohydrate

Streptococcus 'viridans'

A species name loosely used to describe a group of alpha haemolytic streptococci which form part of the normal flora on mucous membranes and which are the usual cause of endocarditis.

Strep. sanguis: commonest cause of endocarditis

Strep. mitis: pyridoxine-dependant strains grow poorly on usual media

Strep. mutans: microaerophilic, associated with dental plaque

Strep. salivarius: dominates aerobic salivary flora but an unusual cause of endocarditis

Strep. milleri (also *Strep. MG, anginosus, intermedius*):
microaerophilic, pyogenic
Strep. bovis: Lancefield gp D but more sensitive to penicillin than
Strep. faecalis

Streptococcus pneumoniae (pneumococcus)
Lanceolate diplococci distinguished by the presence of a large
capsule which has the following characteristics:
a. Composed of carbohydrate, 86 antigenic types
b. Identified by the capsule swelling (quellung) reaction in which
   combination of the capsule with specific antibody causes the
   capsule to become more refractile and appear to swell
c. Essential for pathogenicity. Non-capsulated mutants are
   non-pathogenic
d. Immunity is type-specific. Polyvalent vaccines are available
e. Can be recognised in Gram stains of respiratory secretions as a
   non-staining halo around the diplococci
f. Some capsular types (especially type 3) are associated with
   higher mortality
   Strep. pneumoniae is a normal inhabitant of the upper respiratory
tract. Carrier rates are higher in children.
   Extremely sensitive to penicillin but isolates with reduced
sensitivity have been detected in some parts of the world
(e.g. S. Africa, Papua New Guinea) in recent years.

Spectrum of clinical disease
1. Respiratory tract infection
   Sinusitis, otitis media, conjunctivitis, bronchitis
2. Pneumonia
   Pneumococcus accounts for about 90% of all community-
   acquired lobar pneumonia. Coryza often precedes.
   Together with *Haemophilus influenzae* causes
   bronchopneumonia, especially in patients with chronic lung
   disease.
3. Invasive
   a. Meningitis
   b. Septic arthritis, osteomyelitis
   c. Endocarditis
   d. Peritonitis

High-risk groups:
1. Splenectomy
2. Sickle cell disease
3. Nephrotic syndrome
4. Alcoholics
5. Agammaglobulinaemia

Gram-negative cocci

Two genera are of medical interest (Table 14)

Table 14

	Neisseria	Branhamella
Oxidase test	Positive	Positive
Nutritional requirements	Complex	Simple
Acid from carbohydrates	Yes	No
Growth at 22^{0}C	Variable	No
CO_2 requirement	Variable	No
Habitat	Upper respiratory and genital tracts	Upper respiratory tract
Pathogenicity	Meningitis Gonorrhoea Septicaemia	Occasional cause of bronchitis, sinusitis and otitis media

GENUS NEISSERIA

Includes pathogenic and commensal species. Isolation and identification require attention to the following principles:
1. Neisseria do not resist drying or antiseptics. Specimens must be collected on serum-coated or charcoal swabs and placed in transport medium,
2. Growth requirements:
    a. Multiple amino acids (provided as serum or blood),
    b. Toxic fatty acids must be neutralised with starch, albumin or charcoal
    c. Iron or haeme is required
    d. Optimal atmosphere is 5–10% CO_2 at relative humidity of 70%,

3. Addition of antibiotics is necessary to detect Neisseria at sites of mixed flora (female genital tract, upper respiratory tract). Media can be made selective with vancomycin, trimethoprim, colistin, polymyxin, nystatin,
4. Identification is based on:
   — production of acid from carbohydrates, or
   — antigenic structure of cell wall detected by agglutination

Table 15

	Gonorrhoeae	Meningitidis	Sicca pharyngis	Mucosa	Lactamica
Acid from					
Glucose	+	+	+	+	+
Maltose	−	+	+	+	+
Sucrose	−	−	+	+	−
Lactose	−	−	−	−	+
Colonial pigmentation	−	−	+	+	+
Growth at 22°C	−	−	+	+	+
Resistance to vancomycin	+ (most strains)	+	−	−	−

Pathogenic neisseria

The only pathogenic members of the genus are:
1. N. gonorrhoeae, causing gonorrhoea
2. N. meningitidis, causing meningococcal meningitis
(See Table 15)

Table 16 A comparison of the pathogenic species of Neisseria

	N. gonorrhoea (gonococcus)	N. meningitidis (meningococcus)
Nutritional requirements	Many and variable	5 amino acids
CO₂ requirement	Strict	Variable
Surface antigens		
a. Polysaccharide capsule	?	4 main types A, B, C, D, plus 5 more recent types
b. Pili	Present (correlates with virulence)	?
c. Outer membrane proteins	Antigenically variable	Numerous serotypes (12)

Table 16 Contd

	N. gonorrhoea (gonococcus)	N. meningitidis (meningococcus)
Typing methods	Amino acid requirement ('auxotyping')	Capsular types
Colonial variation	Subculture produces various colonial (Kellog) types associated with loss of pili	Subculture produces rough colonies
IgA protease	Present	Present
Acquired antibiotic resistance	Sulphonamides Penicillins Tetracyclines	Sulphonamides
Betalactamase plasmid	Present in 1–20% of strains	Not detected
Habitat	Genital tract, occasionally pharynx	Pharynx, occasionally genital tract
Carrier state	Inapparent infection in up to 20% of exposed females, 5% of exposed males	Detected in pharynx of 0–60% of children and military recruits
Transmission	Venereal	Respiratory
Systemic infection	Rare (< 1%)	Common
Sites of systemic localisation	Joints, skin	Meninges, skin
Mortality of systemic disease	Low	High
Immunity	Infection does not protect from reinfection	Carrier state elicits immunity
Vaccines	Not available	Available for types A and C. Type B is only weakly immunogenic
Treatment	Penicillin Ampicillin Tetracycline Spectinomycin	Penicillin (sulphonamides or rifampicin for carrier state

Gram-negative bacilli

A diverse and imperfectly-classified group of bacteria having in common only the cell wall structure associated with the Gram-negative staining reaction (see Table 17).

The major classes of facultative anaerobic or microaerophilic Gram-negative rods are as follows:

Enterobacteriaceae: Dominant in the human gut. Some species cause enterocolitis

Pseudomonads: Free-living, may contaminate hospital disinfectants

Vibrios: Free-living, halophilic. *V. cholerae* causes cholera

'Parvobacteria': A term loosely used to describe small, often nutritionally fastidious gram negative rods. Includes several pathogens, both human and animal

Campylobacter: Spirally-curved, microaerophilic rods. Cause enterocolitis in animals and man

Legionellae: A recently described group of fastidious, microaerophilic and yet environmental Gram-negative rods causing pneumonia

ENTEROBACTERIACEAE

A well-defined family of facultative anaerobic, oxidase negative, catalase positive, nitrate-reducing gram negative rods.
— Both free-living and associated with the gut of man and animals
— Simple growth requirements, grow on bile-containing media
The genera are arranged in five taxonomic groups as follows:

  I *Escherichia*
    Dominant in the human gut. Commonest cause of renal tract infection and septicaemia
    Shigella
    Important cause of enterocolitis
 II *Citrobacter*
    Hospital opportunist
    Salmonella
    Important cause of enterocolitis

Table 17 Characteristics of major groups of Gram-negative rods

	Morphology	Motility	Metabolism	Oxidase	Growth requirements	Habitat	Infections
Enterobacteriaceae	Straight rods	Peritrichous flagella	Respiratory and fermentative Acid from glucose	Negative	Simple	Gut and free-living	Enterocolitis Peritonitis Pyelonephritis
Pseudomonads	Straight rods	Polar flagella	Respiratory Strict aerobes	Usually positive	Very simple	Free-living Wide temperature range.	Opportunistic. Nosocomial
Vibrios	Curved rods	Single polar Darting	Respiratory and fermentative	Positive	Simple, with better growth above pH 7	Free-living Halophilic Alkaline	Enterocolitis
"Parvobacteria"	Short rods 0.3 × 2.0 μm	Mostly non-motile	Variable	Variable	Usually complex	Pathogens of animals and man	Various
Campylobacter	Spirally-curved rods	Single polar flagellum	Respiratory Carbohydrates not utilised	Positive	Complex	Animals	Enterocolitis
Legionellae	Short rods	Motile	Respiratory	Variable	Complex	Water supplies	Pneumonia

III *Klebsiella*
   Enterobacter
   Serratia
   Free-living, utilise simple carbon-containg compounds for growth. Some are pigmented or capsulated. Fermentation produces acetoin. Hospital opportunists
IV *Proteus*
   Providence
   Morganella
   Deaminate amino acids. Swarm on agar. Cause renal tract infection, sometimes with stone formation because of ability to split urea
 V *Yersinia*
   Primarily cause diseases in animals; plague and yersiniosis in man

Surface components and antigens of Enterobacteriaceae (see Table 18).

1. Flagella
— composed of contractile fibrous proteins (flagellins)
— constitute the H antigens (H = *Hauch* (German) = breath)
— preserved by formalin, destroyed by heat
— maximal in young cultures

2. Fimbriae (pili)
— hairlike projections of protein on cell surface, promote adhesion
— develope in old (24–48 h) broth cultures
— destroyed by heat, preserved by formalin
— six types are described, according to ability to agglutinate red cells, inhibition by mannose, width
— may interfere with H agglutination

3. Capsule
— a thin layer of surface polysaccharide constitutes the K antigen (K = *Kapseal* (German) = capsule)
— formation is enhanced by sugar-containing media
— very prominent in klebsiellae, in which as a loosely adherent slime layer, it creates mucoid colonies
— may interfere with 0 agglutination
— inhibits phagocytosis
— may cross react with capsular antigens of other bacteria

4. Cell wall.
Consists of two layers:
a. Outer layer
   A complex of lipopolysaccharide (LPS), protein and lipids which render Gram-negative cells less permeable to solutes.

Table 18 Structural Components and Antigens of some of the Enterobacteriaceae

Component	Esch. coli	Shigella	Salmonella	Klebsiella
Flagella (H antigen)	More than 49 types	Nil (shigellae are non-motile)	Spontaneous variation between 2 antigenic phases designated 'specific' (phase I) and 'non-specific' (phase II)	Nil (klebsiellae are non-motile)
Fimbriae 3 types	Type I fimbriae 3 types Widely shared antigens	Only found on Sh. flexneri. Common antigen	Type I fimbrial antigen shared by most strains	One or more types present on most strains
Capsule (K antigen)	More than 190 types comprising 3 types of antigen (A, L and B) usually dependent on O type	K antigens specific for each serotype	Vi (virulence) antigen of Salm. typhi may hinder O agglutination M antigen of Salm. paratyphi B.	72 capsular types. Antigen is heat stable and masks O antigen
Outer cell wall polysaccharide (O antigen)	More than 140 types	dysenteriae: 10 types flexneri: 6 types plus 8 shared group antigens. boydii: 15 types sonnei: one antigen with rough variant.	More than 60 types, some determined by presence of lysogenic phage	5 types, related to Esch. coli
Clinical significance of antigenic structures	Certain O types are associated with infant diarrhoea, pyelonephritis Certain K types are invasive (e.g. neonatal meningitis)	Used epidemiologically	O and H antigens are used to create more than 1000 serotypes (Kauffman–White scheme), each given a species name. Used to trace food-borne epidemics Vi antigen is present on Salm. typhi	Some capsular types cause pneumonia

   (i) The *side chains* of repeating sugar units project from the outer LPS layer constituting the O antigen (O = *Ohne* (German) = without). Associated with smooth colonies, resistance to killing by complement. Absence of side chains is associated with rough colonies, autoagglutination, non-virulence and killing by complement

  (ii) *Core glycolipids* form the basal layer to which side chains are attached (the enterobacterial common antigen)

 (iii) *Lipid A* forms a layer between the lipopolysaccharide and a phospholipid membrane. Is the toxic moiety of 'endotoxin'

 (iv) *Phospholipid membrane* similar in structure to the cell membrane (and therefore termed the 'outer membrane')

  (v) *Proteins* (outer membrane proteins, OMPs) are present in the phospholipid membrane. They include those responsible for solute transport ('porins') and structural lipoproteins

5. *Inner layer of cell wall*
— consists of peptidoglycan maintaining cell wall rigidity

6. *Exotoxins*
a. enterotoxins: produced by some strains of *Esch. coli* and probably other Enterobacteriaceae.
b. cytotoxins: *Sh. dysenteriae* and some strains of *Esch. coli*.

   Classification within the family is based on a collection of tests commonly employed in medical laboratories (see Table 19).
  1. Fermentation of lactose
    — distinguishes two genera causing enterocolitis (*Shigella* and *Salmonella*) from bowel commensals
    — lactose plus indicator are incorporated into commonly MacConkeys, deoxy ... used selective media (e.g MacConkey's, deoxycholate citric agar (DCA)).
  2. ONPG (o-nitrophenyl-B-D-galactopyranoside)
    — hastens the detection of potential lactose fermenters because ONPG enters cells without the aid of permease and is then hydrolysed by B-galactosidase to the yellow o-nitrophenol
  3. Fermentation of other carbohydrates
    — numerous monosaccharides and disaccharides are used, mostly to distinguish species within a genus
  4. Gas production from glucose
  5. Motility
    — is a time-consuming test not commonly used in primary identification.
    — distinguishes two non-motile genera (Shigella and Klebsiella).
  7. H2S production

Table 19 Major identifying characteristics of Enterobacteriaceae (key reactions are circled, d = 20–80% positive)

	Motility	Lactose	Sucrose	ONPG	Tryptophan deaminase	Urease	Indole	H_2S	Citrate	Acetoin	Gelatin	Gas in glucose	Arginine	Lysine	Ornithine
Esch. coli	+	⊕	d	+	−	−	⊕	−	−	−	−	+	d	+	+
Shigella	⊖	⊖	−	d	−	−	d	−	−	−	−	−	−	−	−
Citrobacter	+	+	+	+	−	d	d	⊕	+	−	−	+	d	−	+
Salmonella enteritidis	+	⊖	−	⊖	−	−	−	⊕	+	−	−	+	+	+	+
Salmonella typhi	+	⊖	−	⊖	−	−	−	⊕	−	−	−	−	−	+	−
Klebsiella aerogenes	⊖	+	+	+	−	+	−	−	+	⊕	−	+	−	+	−
Enterobacter species	+	+	+	+	−	+	−	−	+	⊕	d	+	+	+	+
Serratia species	+	+	+	+	−	d	−	−	+	+	⊕	d	−	+	+
Proteus species	+	⊖	d	⊖	⊕	⊕	d	⊕	d	d	d	d	−	−	d
Providencia species	+	⊖	d	⊖	⊕	−	d	−	+	−	−	d	−	−	−
Yersinia	+	+	d	+	−	d	d	−	−	d	−	−	−	−	d

8. Acetoin (Voges-Proskauer test)
   — is a characteristic fermentation product of klebsiellae.
9. Deamination of phenylalanine or tryptophan
   — is characteristic of Proteus and Providentia species.
10. Decarboxyalation of arginine, lysine and ornithine.
    — distinguishes species within genera.
11. Urease production
    — is characteristic of Proteus species.
12. Production of indole from tryptophan
13. Hydrolysis of gelatin and DNA

Selective agents used for Enterobacteriaceae (see also Table 21)

Table 20

Agent	Selects	Inhibits
Bile (5%) Na taurocholate	Enterobacteriaceae prevents swarming	Gram-positives
Dyes (eosin, methylene blue)	Enterobacteriaceae	Gram-positives
Deoxycholate	Salmonella Shigella Proteus	Esch. coli Klebsiella
Citrate	Salmonella	Esch. coli
Brilliant green Bismuth sulphite	Salmonella	Other Enterobacteriaceae
Tetrathionate Selenite	Used in liquid media to select Salmonella	Esch. coli, Klebsiella, Shigella

Table 21 Selective media for Gram-negative bacteria

Medium	Constituents	Purpose
MacConkey agar	Taurocholate (bile) Lactose Neutral red	Bile inhibits swarming of Proteus and is selective for coliforms. Lactose-fermenting colonies are pink
	Crystal violet (optional)	Inhibits Gram-positives
Deoxycholate agar (DCA)	Deoxycholate Fe^{+++} amm. citrate Thiosulphate Lactose Neutral red	Selects Salmonella, Shigella and Proteus Inhibits Esch. coli and Klebsiella. Lactose-fermenting colonies are pink. H_2S production causes blackening

Table 21 Contd

Medium	Constituents	Purpose
XLD	Xylose Sucrose Lactose Lysine Thiosulphate Fe^{+++} amm. citrate Deoxycholate Phenol red	Helps eliminate *Proteus* species (ferment xylose, do not decarboxylate lysine)
Bismuth sulphite agar	Peptone/agar Glucose $Fe^{++} SO_4$ Bismuth amm. citrate Na suphite Brilliant green	Highly selective for salmonellae, especially *Salm. typhi*
Triple sugar iron agar (TSI)	Glucose Lactose Sucrose	Used as a screening test for non-lactose fermenting isolates in stabbed agar slants. Detects fermentation of 3 sugars, gas production, H_2S production
Tetrathionate broth	Thiosulphate Potassium iodide Ca carbonate Phenol red	Used as an enrichment medium. Inhibits coliforms at the expense of *Salmonella* and *Proteus*
Selenite broth	Na selenite	As tetrathionate broth
Thiosulphate citrate bile sucrose agar (TCBS)	NaCl 10% Sucrose Citrate Thiosulphate Bile Bromthymol blue pH 8.6	High salt content, high pH. Sucrose replaces lactose. Selective for *Vibrio* species
Alkaline peptone water	Peptone water, pH 8.4	Selects *Vibrio* species
Cetrimide agar	Peptone/agar Cetrimide	Selects pseudomonads (which tolerate quaternary ammonium compounds)
CIN agar	Peptone Mannitol Pyruvate Deoxycholate 'Irgasan' Neutral red Crystal violet Cefsulodin	Selective for *Yersinia enterocolitica*

Table 21 Contd

Medium	Constituents	Purpose
Campylobacter medium	Blood agar Vancomycin Polymyxin B Trimethoprim Amphotericin B Cephalothin	Selective for *C. fetus*, ss *jejuni* (at 42°C in 5% oxygen and 10% CO_2).
BCYE	Yeast extract Fe^{+++} pyrophosphate ACES buffer, pH 6.9 Alpha-ketoglutarate Activated charcoal L-cysteine Cefamandole Polymyxin Anisomycin Vancomycin	A complicated medium for growing and selecting *Legionella* species at optimal pH

Diseases produced by enterobacteriaceae
1. *Enterocolitis*
   a. *Salmonella*
   b. *Shigella*
   c. *Esch. coli*
      Enterotoxigenic—in travellers
      Enteropathogenic—in nurseries
      Invasive
   d. *Yersinia enterocolitica* and *pseudotuberculosis*
2. *Renal tract infection*
3. Infection following *anatomical damage of bowel* and related organs (usually combined with anaerobes)
   — Peritonitis, diverticulitis, appendicitis
   — Cholangitis
   — Endometritis
4. *Septicaemia*, following 2. or 3.
5. *Invasive disease*, due to certain capsular types:
   Esch. coli—Capsular type K1—Meningitis in neonates
   Salm. typhi—Vi antigen—Enteric (typhoid) fever
   K. aerogenes—'respiratory types'—Friedlander's pneumonia

PSEUDOMONAS SPECIES

A group of strictly aerobic, usually oxidase positive rods widely distributed in the environment, nutritionally versatile and tolerant of dilute disinfectants (see also Table 22).

Table 22 Characteristics of some free-living Gram-negative bacilli

	Morphology	Metabolism	Oxidase	Habitat	Infections
Pseudomonas species	Polar flagella	Respiratory Obligate aerobes	Positive	Water supplies Many tolerate detergents	*P. aeruginosa* — opportunist *P. pseudomallei* — melioidosis *P. mallei* — glanders
Acinetobacter calcoaceticus	Coccobacilli (diplococci) Non-motile	Oxidative	Negative	Indigenous flora of skin and mucosa	Colonises airways following intubation, occasionally causes pneumonia, venous line bacteraemia
Aeromonas liquefasciens (hydrophila)	Polar flagella	Facultative	Positive	Salt and fresh water	Secretory diarrhoea Sepsis in immunosuppressed
Aeromonas salmonicida	Non-motile	Facultative	Positive	Cold-blooded animals	
Pleisiomonas shigelloides	Polar flagella	Facultative	Positive	Water sources	Diarrhoea? mechanism
Vibrio species	Curved rods Single polar flagellum	Facultative	Positive	Surface and marine waters	Secretory diarrhoea Cellulitis
Alcaligenes species	Peritrichous flagella	Oxidative	Positive	Water and soil	Contaminate wounds, urine, sputum Rarely invasive
Flavobacterium species	Non-motile	Weakly fermentative	Positive	Water and soil	*F. meningosepticum* is an occasional cause of meningitis in neonates
Chromobacterium violaceum	Polar and Lateral	Facultative	Variable	Water and soil	Systemic infections in compromised hosts

— Cause infections in compromised hosts
— Classified according to:
  a. flagella number
  b. pigment production
  c. utilisation of various substrates
— Major species include:
  P. aeruginosa
    Important hospital opportunist
  P. cepacia
  P. maltophilia
  P. putida
  P. fluorescens
    Hospital contaminants
    Occasionally isolated from
    disinfectants, wounds, sites
    of venous cannulation, bladder
  P. pseudomallei
    Causes melioidosis
  P. mallei
    Causes glanders in horses, occasionally in man

Pseudomonas aeruginosa
— Produces large, usually rough colonies, characteristic smell
— Characteristic blue-green pigment surrounds colony
— Secretes numerous compounds into surrounding medium:
  Proteolytic enzymes
  Phospholipases, lecithinase
  Pigments, including pyocyanin (blue-green)
  Exotoxins
  Pyocins (bacteriocins)
— Found in water, soil, fruit
— Colonisation rate rises during hospitalisation (from about 5% to 20%)
— Causes disease in the presence of host defect, as follows:
  Neutropenia: septicaemia
  Premature neonates: septicaemia
  Cystic fibrosis: Bronchopneumonia
  Tracheostomy, intubation: pneumonia
  Burns: Septicaemia
  Drug addicts: Endocarditis
  Catheterisation: Urinary tract infection
  Surgical wounds: Wound infection

VIBRIOS

A large and rather poorly-characterised group of Gram-negative rods, mainly environmental, especially in marine and brackish waters. Tolerate high salt concentrations (i.e. halophilic). Curved rods, single polar flagellum. Oxidase positive, catalase positive. Fermentative.
— Species of medical interest

V. cholerae
Includes 72 serotypes, one of which (01) causes epidemic cholera

V. parahaemolyticus
Causes gastroenteritis (from seafood)

V. alginolyticus
Occasionally causes wound infection following wounds sustained in seawater

V. vulnificus
Occasionally causes septicaemia after eating raw oysters

Vibrio group F(EF-6)
Has been recently described as a cause of gastroenteritis

V. cholerae

— Grows poorly on media selective for *Salmonella* and *Shigella*
— Produces colourless colonies on MacConkey at 18–24 h
— Is best isolated using TCBS medium, with or without prior enrichment in alkaline peptone water
— Biochemical characteristics include: lactose −, indole +, gelatinase +, arginine −, lysine +, ornithine +
— Includes 72 0 serotypes, only one of which (01) causes epidemic cholera. Other serotypes, known as non-01 or 'non-agglutinable' (NAG) vibrios, do not cause cholera
— Includes two biotypes of serotype 01 with the following distinguishing features:

	Classical	Eltor
Acetoin production	+	−
Polymyxin	sensitive	resistant
Carrier state	+/−	++
Survival in surface waters	+	++

— Colonisation of small bowel and production of enterotoxin causes watery diarrhoea
— Successive pandemics have spread to the West from India
— Despite severe disease during epidemics, mild or asymptomatic cases predominate
— Most disease is transmitted by water in which *V. cholerae* can survive for long periods, especially if alkaline

'PARVOBACTERIA'

The term is used here to include small, nutritionally fastidious gram negative rods of various genera (Table 23).

Table 23

Genus	Habitat	Diseases in man
Haemophilus species	Respiratory tract of man	Respiratory tract infections Capsulated strains are invasive
Bordetella pertussis	Respiratory tract of man	Whooping cough
Brucella species	Urogenital tract of animals	Brucellosis
Francisella tularensis	Animals	Tularaemia
Pasteurella	Animals	Infection of animal bites
Yersinia species	Animals	Plague Mesenteric adenitis Enterocolitis
Eikenella corrodens	Human oral cavity	Pyogenic
Streptobacillus moniliformis	Nasopharynx of rats	Rat-bite fever
Moraxella liquefasciens	Respiratory tract	Conjunctivitis
Actinobacillus	Animals	Endocarditis (rare)

Haemophilus species

— Small, pleomorphic, facultative anaerobic, strictly parasitic
— Usual habitat is respiratory tract
— Growth requirements:
  'V' factor — NAD, NADP or nicotinamide nucleoside
  'X' factor — haematin, porphobilinogen (not required anaerobically)
  Media used for growing haemophili include
— Blood (but note that sheep blood inhibits)
— Chocolate blood (releases V factor)
— Filde's digest (peptic digest of blood)
— Lysed blood
— Levinthal's (5% heat-lysed blood, clear medium)
  Initial classification (Table 24) is based on growth requirements and haemolysis:

Table 24

	X	V	Haemolysis	Disease
H. influenzae	+	+	−	Respiratory and invasive infections
H. parainfluenzae	−	+	−	Less pathogenic than H. influenzae
H. haemolyticus	+	+	+	
H. parahaemolyticus	−	+	+	
H. aegyptius	+	+	−	conjunctivitis
H. ducreyi	+	−	+/−	chancroid
H. aphrophilus	+	−	−	endocarditis (rare)

Haemophilus influenzae
— An important cause of respiratory tract infections
— Encapsulated strains may be invasive in the first years of life
— Coccobacillary in young cultures, larger bacillary and filamentous forms in old cultures and purulent spinal fluid
— Capsulated strains produce larger, more opaque colonies, iridescent in oblique light (on a transparent medium)
— Six capsular types, designated a–f (Pittman) are detected by quellung reaction or agglutination. Each is a carbohydrate polymer. Type b is polyribose phosphate.
— Capsular type b causes most invasive disease with peak incidence between 3 months and 5 years of age. Clinical manifestations:
1. Nonencapsulated strains cause sinusitis, otitis media, bronchitis and bronchopneumonia in predisposed individuals.
2. Encapsulated strains, usually gp b, are invasive in the first years of life, causing:
   — Epiglottitis
   — Pneumonia, pericarditis
   — Cellulitis — cheek, periorbital
   — Meningitis
   — Septic arthritis
Treatment
1. Ampicillin, amoxycillin (but note that some strains produce betalactamase)
2. Chloramphenicol — especially for meningitis
3. Trimethoprim compound
4. New cephalosporins
5. Rifampicin — for respiratory carriers of type b

Bordetella species
Very small (<1 μm) ovoid bacilli, including 3 species (pertussis, parapertussis, bronchiseptica). All share a common O antigen.

B. pertussis
— The usual cause of whooping cough
— Requires enriched medium containing catalase and a substance
  to absorb toxic fatty acids (e.g. albumin, charcoal, starch, blood)
— Bordet–Gengou medium (potato-glycerol agar plus 33% blood)
  is commonly used. Cephalexin improves selectivity
— Whitish, refractile colonies ('bisected pearls') appear in 2–3 days
— Virulent strains have capsules which include a dominant antigen
  common to the species (used as an identification test on
  colonies). Additional antigens allow serotyping
— Produce toxins:
   (i) Heat labile (mouse lethal)
   (ii) Lymphocytosis-stimulating factor

Whooping cough (pertussis)
Incubation period 7–10 days. Catarrh followed by characteristic
paroxysmal cough and lymphocytosis. Prolonged convalescence.
Secondary bacterial infection causes otitis media, pneumonia.
Severity is greatest in infants less than 6 months.
   Erythromycin, tetracycline and chloramphenicol (not ampicillin)
eliminate the organism but do not aleviate the cough. Steroids may
ameliorate the paroxysms.
   Vaccines give 80% protection in the first 3 years after
administration.

Brucella species
Causes infections in domestic animals. Incidentally passed on to
man.
   Round to oval coccobacilli. Small transparent colonies, may take
several days to develope. *B. abortus* has a variable requirement for
CO_2. Six species, classified according to CO_2 requirement, H_2S and
urease production, inhibition by dyes (Table 25).

Table 25

	Host	Pathogenicity, distribution
B. abortus	Cattle	Most common species in UK, Australia
B. melitensis	Goats	Mediterranean, Arab peninsula
B. suis	Pigs	Meat packers, USA
B. canis	Dogs	Rare
B. ovis	Sheep	Do not cause disease in man
B. neotomae	Woodrats	

B. abortus, melitensis and *suis* have a close antigenic relationship. Cause bacteraemia in animals, later settles in seminal vesicles, uterus, mammary glands.
Transmitted to man by:
a. Drinking unpasteurised milk, cheese, butter
b. Inhalation (laboratory)
c. Occupational exposure — abbatoirs, farmers, veterinarians.

Clinical manifestations
1. *Acute: fever 7–21 days after exposure, backache, arthralgia, weight loss, splenomegaly*
2. *Localised: especially B. suis and melitensis.* Lung, bones, cardiac valves, meningitis, testis
3. Relapses: may occur up to 3 months after onset
4. 'Chronic': The existence of this entity is disputed

Yersinia species
— Small, pleomorphic bipolar-staining rods. Oxidase negative
— Small colonies after 24 hours
  Three species are of medical interest (Table 26)

Table 26

Characteristic	Y. pestis	Y. enterocolitica	Y. pseudotuberculosis
Motility	−	+	+
Urease	−	+	+
Acetoin production	−	+	−
Habitat	Small mammals	Various domestic animals	Wild and domestic mammals and birds
Vehicle of transmission	Fleas	Water, food, milk	Water, food, milk
Clinical features	Plague	Enterocolitis Mesenteric adenitis Reactive arthritis	Mesenteric adenitis

Plague
— Is normally a disease of wild rodents (sylvatic plague), occasionally transmitted from them to rats and humans by fleas
— Has caused three human pandemics since the sixth century AD but currently causes only sporadic disease in hunters who capture and skin small carnivores
— Surface structure of *Y. pestis* includes a protein (Fraction I) which inhibits phagocytosis

— Produces two clinical pictures:
  a. Bubonic plague
     Follows 2–6 days after a flea bite. Regional lymph node enlargement (bubo), endotoxaemia with vascular damage, haemorrhagic necrosis, pneumonia
  b. Pneumonic plague
     Follows inhalation of droplets from patients with the disease. Short incubation period (about 2 days) and rapid death
  Plague can be treated with tetracyclines, streptomycin, chloramphenicol, sulphonamides, trimethoprim. The patient's contacts should also be treated.
  A formalin-killed whole organism vaccine is used for persons working in high-risk occupations.

Yersiniosis
The term is used for the manifestations of infection due to *Y. enterocolitica*, seen mostly in the cooler regions of Europe and North America, causing:
a. Enterocolitis: young children
b. Mesenteric adenitis: older children. Mimics appendicitis
c. Septicaemia: especially in patients with iron overload receiving desferrioxamine (*Y. enterocolitica* is exceedingly iron-dependent)
d. Autoimmune diseases: especially in women. Includes erythema nodosum and reactive arthritis

CAMPYLOBACTER

Somewhat fastidious pathogens and commensals of man and animals. Thin curved rods, occasionally S-shaped, 'gull-wings' or long spirals. Single polar flagellum causes darting motility. Oxidase positive. Sugars are not oxidised or fermented. Some species are microaerophilic (grow only at O_2 concentrations of 3–15%).
   After the introduction of a selective medium in 1977, *C. fetus* subsp. *jejuni* was revealed to be the most common bacterial cause of enteritis in man.
   The species of Campylobacter are divided into subspecies as follows:
1. *C. fetus*
   subsp. *fetus*: Causes abortion and sterility in cattle
   subsp. *intestinalis*: Outbreaks of abortion in sheep. Occasional cause of septicaemia in compromised human hosts
   subsp. *jejuni*: Common cause of enteritis in man and domestic animals
2. *C. sputorum*
   subsp. *sputorum*: Normal human oral flora
   subsp. *mucosalis*: Enteritis of pigs
3. *C. faecalis*: Commensal of sheep and cattle
4. *Campylobacter*-like organisms: cause proctitis in male homosexuals. Taxonomic status uncertain.

LEGIONELLAE

The genus was established following the discovery of a Gram-negative bacillus which caused an outbreak of pneumonia amongst American Legionnaires attending a convention in Philadelphia in 1976.

— Small, pleomorphic rods which stain poorly with gram stain but well with silver
— Catalase positive, oxidase variable, motile (single polar flagellum)
— Cell walls contain unique, branched fatty acids
— No growth on blood agar. Slow growth (4–5 days) on media supplemented with cysteine, charcoal, haeme. Growth is optimal at pH 6.9 in 5% CO_2. Strictly aerobic.
— Colonies fluoresce in ultraviolet light
— Multiplies in macrophages. Cellular immunity is important for recovery
— Found in water supplies in which it probably multiplies in freeliving blue-green algae and amoebae. Survives temperatures of 6–67°C
— An increasing number of species is being defined, each with a distinct fatty acid composition

L. pneumophila
6 serotypes based on 0 antigens. Type I causes most pneumonia. Produces soluble brown pigment

L. micdadei
Weakly acid fast. Occasional cause of opportunistic pneumonia

L. bozemanii
L. dumoffii
L. longbeachae
L. gormanii
L. jordanis
Occasional reports of opportunistic pneumonia

Legionnaires' disease
— Is mostly caused by L. pneumophila, type 1
— Accounts for about 5% of community acquired pneumonia
— When occurring in outbreaks has been associated with air conditioning, cooling towers, excavations
— When causing sporadic disease tends to occur in late summer
— Occurs more commonly in elderly males with smoking-related bronchitis. Opportunistic in immunosuppressed patients
— Is diagnosed by:
  a. Detection of the organism in bronchoscopically-obtained respiratory secretions or lung biopsy using immunofluorescent antibodies or culture
  b. demonstration of a rising titre of antibodies (usually immunofluorescent). May take many days.

Gram-positive rods

Table 27 Gram-positive rods of medical interest

Genus	Morphology	Diseases produced
Corynebacterium	Club-shaped	Mostly commensals except *C. dipheriae*
Nocardia	Branching, filamentous Weakly acid fast	Cause opportunistic disease
Streptomyces	Filamentous	Non-pathogenic. Source of several antibiotics
Actinomyces	Filamentous Anaerobic	Rare cause of infections relating to oral cavity, large bowel, female genital tract and lung
Mycobacterium	Acid fast	Tuberculosis
Listeria	Short, motile	Listeriosis
Erysipelothrix	Short rods in chains	Erysipeloid
Bacillus	Large, spore-forming	Soil organisms Anthrax, diarrhoea
Gardnerella	Small, Gram-variable	Vaginal commensal associated with non-specific vaginitis

CORYNEBACTERIA

A genus of diverse animal and plant pathogens and commensals.
— Straight or slightly curved rods, some with club-shaped terminal swellings (*corynos* (Greek) = a club) and irregularly stained segments
— May contain volutin granules which are deposits of polymerised metaphosphates, more plentiful in older cultures, stain metachromatically

— 'Snapping' division produces angular ('Chinese character') or pallisade arrangements
— Catalase positive
— Cell wall composition is similar to mycobacteria and nocardiae but with mycolic acids of shorter chain length and therefore not acid-fast
— Growth of pathogenic species is improved by blood or serum
— Resistant to 0.03% tellurite, which is used as a selective isolation medium (e.g. Hoyle's, McLeod's) but note that other organisms which grow on this medium include diphtheroids, *Streptococcus faecalis*, *Erysipelothrix rhusiopathiae* and *Listeria monocytogenes*

Identification methods for corynebacteria
1. Morphology
   — is best determined on colonies grown on Loeffler's inspissated serum slopes using metachromatic stains to detect granules
2. Colonial appearance on tellurite-containing media
3. Reduction of nitrate
4. Fermentation of carbohydrates
5. Haemolysis on blood agar
6. Urease production
7. Toxin production (*C. diphtheriae*)

Major species of corynebacteria
C. diptheriae (see below)

C. ulcerans: occasional cause of exudative pharyngitis some strains produce toxins

C. xerosis
C. hofmanii   human skin commensals
Corynebacterium, group JK

C. pyogenes   Cell walls, fermentation products and
C. haemolyticum   serological relationships resemble streptococci and actinomycetes. Catalase negative, betahaemolytic

C. bovis   Animal pathogens. Occasionally cause human
C. ovis   lymphadenopathy, pulmonary disease.
C. equi

C. minutissimum: causes erythrasma

Corynebacterium diphtheriae
— Isolated by Loeffler in 1884
— Multiplication of toxigenic strains on tonsils or posterior pharynx causes elaboration of exotoxin, necrosis of tissues and formation of pseudomembrane
— May extend to palate or larynx and trachea, causing airways obstruction
— Absorption of toxin causes neuritis (especially palatal paralysis) and carditis
— Three named biotypes of *C. diphtheriae* were previously related to the severity of the disease but are still useful epidemiologically (see Table 28)

Epidemiology
— Acquired by inhalation of droplets from patients or carriers
— In tropical countries, skin infections due to *C. diphtheriae* or *C. ulcerans* in young children may confer immunity
— Carrier rates in developed countries are low

C. diphtheriae toxin
— was discovered by Rous and Yersin in 1888
— is a protein of mol. wt. 64 000 with 3 or 4 distinct antigens
— binds to specific receptors on tissue cells, inhibiting protein synthesis by interfering with transfer RNA
— produced only by strains containing a lysogenised bacteriophage
— is produced maximally in young dividing cells, in low Fe^{++} concentration
— is inactivated by formalin but retains immunogenicity (toxoid). Used for immunisation
— when injected intradermally in a small dose causes erythema formation in 24 hours (Schick test). Previous immunisation prevents the reaction
— is detected in cultures of the organism by
   a. in vitro: precipitation in agar (Elek plate)
   b. in vivo: in guinea pigs, as follows:
      (i) intraperitoneal injection, causing death
      (ii) intradermal injection, causing skin necrosis

Treatment
— should be commenced as soon as the clinical diagnosis is made (*not* when confirmed by laboratory culture)
— give 1000–100 000 units of antitoxin i.m.
— eliminate pharyngeal carriage of the organism with erythromycin

'Diphtheroids'
A loosely-used term implying corynebacteria other than *C. diphtheriae*. General differences between *C. diphtheriae* and diphtheroids are given in Table 29.

Table 28 Characteristics of the three biotypes of *C. diphtheriae*

	Gravis	Intermedius	Mitis
Microscopy	Short Uniform	Long Barred Some clubbing	Long Curved Cuneiform arrangements
Pleomorphism	+	+ +	+ + +
Granules	+/−	+/−	+ +
Barring	+	+	+/−
Colonies (on tellurite agar)			
18 hours	1–2 mm Circular Low Convex Pearly grey or blackish centre	< 1 mm Uniform, small discrete Sl. raised Sl. umbonate Smooth	< 1.0–1.5 mm Circular, convex Mushroom grey Entire edge Smooth, glistening
2–3 days	3–5 mm Crenated edge Radial striation 'Daisy head'	Dark centre Light edge 'Frog egg'	2–4 mm 'Poached egg'
Consistency	Brittle and coherent 'Cold margarine'	Intermediate	 'Soft butter'
Haemolysis	Some strains	No	Weak beta
Toxin production, pathogenicity	+ + +	+ +	+
Growth on blood agar	Large, convex	Flat, creamy	Large, convex
Fermentation	Starch, glycogen		
Ultrastructure	3 layers	2 layers	3 layers
Wall thickness	+	+ + +	+
mesosomes	+	+ + +	+ + +
K antigens (serotypes)	13	14	40–50
		Lipophilic Poor growth on simple media	

Table 29

	C. diphtheriae	Diphtheroids
Arrangement	Chinese letters	Palisades
Staining	Irregular	Regular
Shape and size	Irregular	Regular
Fermentation of sucrose, urease	Always negative	May be positive
Granules	Prominent	Rare
Appearance on tellurite agar	Irregular edge Differentiated	Entire edge Undifferentiated
Toxin production	Positive	Negative

BACILLUS SPECIES
— Large spore-bearing bacilli. Free-living
— Some species are only Gram-positive in young cultures
  Two species are of clinical interest, viz:
1. B. anthracis
   — causes anthrax
2. B. cereus
   — causes food poisoning and enterotoxigenic enteritis
   — occasionally causes septicaemia in neutropenia patients
   Three species are used to confirm the efficacy of sterilising
procedures, viz:
1. *B. stearothermophilus*—autoclaves
2. *B. subtilis*, var. *globigii*—ethylene oxide
3. *B. pumilis*—ionising radiation

Anthrax
— Caused by B. anthracis, a soil organism causing disease primarily
  in herbivorous animals, especially cows and sheep.
— Large, rectangular rod. Medusa head colonies.
— Surrounded by a capsule of D-glutamylpolypeptide.
— Produces a toxin.
— Must be distinguished from *B. cereus* (Table 30):

Table 30

	B. anthracis	B. cereus
Capsule	+	−
Motility	−	+
Penicillin	Sensitive	Resistant
Lysis by gamma phage	Yes	No
Toxin production	Yes	Yes (weak)
Haemolysis on blood agar	−	+
Lecithinase	−/+	+

Anthrax is largely an occupational disease of farmers, veterinarians, dock workers, and workers with hides and bone-meal.
Two clinical syndromes:
1. Cutaneous: necrotic lesion surrounded by gross oedema
2. Pulmonary: rapid dissemination, often fatal
   Treatment is with penicillin.
   An alum-precipitated culture filtrate can be used to vaccinate workers at high risk.

LISTERIA

Listeria monocytogenes is a short Gram-positive rod, motile at room temperature.
Colonies are surrounded by narrow zones of beta haemolysis.
— Catalase positive, produces acetoin. Hydrolyses aesculin
— The organism and its colonies are easily confused with diphtheroids and streptococci unless properly identified
— The organism is widespread in animals and the environment, growing at temperatures above 4°C

Clinical syndromes
1. Sepsis, without organ localisation
2. Meningoencephalitis
3. Disseminated abscesses and granulomas (neonates)

Factors predisposing to infection:
1. Pregnancy
   — mostly in the third trimester
   — transplacental transmission causes granulomatosis infantiseptica
2. Immunosuppression

ACTINOMYCETACEAE

A family of branched filamentous bacteria related to mycobacteria by cell wall composition.
Includes two groups of medical interest (Table 31).

Table 31

	Aerotolerance	Habitat	Diseases produced
Nocardia	Strictly aerobic	Soil	Pulmonary and disseminated infection, especially in immunosuppressed. Maduramycosis.
Actinomyces	Micro-aerophilic or anaerobic	Oral cavity	Suppurative lesions of face, lung, abdomen often with sinus formation Pelvic infections in IUCD users

Nocardiosis

Three species are involved:

N. asteroides
— 50% of infections occur in patients with immune defect, mostly cell-mediated
— Usually enters via respiratory tract causing pulmonary infiltrates. Cavitation or empyema may follow
— Dissemination particularly involves brain

N. brasiliensis
— Mycetoma in Mexico and S. America

N. caviae
— Mycetoma, occasionally systemic disease

Actinomycosis

— Mostly due to *A. israelii*, occasionally other species
— Cause single or multiple suppurative lesions which track through tissue planes causing sinus formation if skin is breached
— The exudate of sinuses may contain 'sulphur granules', distinctive masses of filamentous gram positive bacteria with a clubbed peripheral fringe of filaments encased in protein
— Often present in mixed culture and may be difficult to isolate
— Sensitive to most antibiotics. Prolonged penicillin is used most commonly
Four sites are involved (Table 32)

Table 32

Site	Prelude	Clinical manifestations
Cervico-facial	Dental infection and manipulation	Tissue induration, cutaneous fistulas
Thoracic	Aspiration of infected oral debris	Non-specific pulmonary opacities, cavitation, empyema, chest wall sinuses and bone destruction
Abdominal	Breach of intestinal mucosa	Mostly ileocaecal following appendiceal surgery. Insidious onset, spreads to abdominal wall, retroperitoneal tissue and pelvis
Pelvic	From intestine or endometrium	Recent cases of chronic pelvic inflammatory disease following insertion of intrauterine device

GARDNERELLA VAGINALIS (previously known as *Haemophilus vaginalis, Corynebacterium vaginale*)

— Gram-variable coccobacilli present in female genital tract
— Small colonies, haemolytic on human blood agar. Oxidase and catalase negative
— Hydrolyses starch and hippurate, moderately sensitive to metronidazole
— Associated with non-specific vaginosis

Anaerobes

Bacteria which will not grow, or which perish, in the presence of oxygen.

Methods for creating anaerobic conditions

1. *Within the medium*
a. Boiling, followed by sealing the tube or flask
b. Deep cultures, sloppy agar, gelatin — reduce the diffusion of oxygen to the depths of the medium
c. Overlay of the medium with paraffin or petroleum jelly
d. Addition of reducing agents: thioglycollate, cysteine, cooked meat, ascorbic acid, sulphites, reduced iron, alkaline glucose

2. *In the atmosphere over agar medium*
a. Roll tubes: agar is coated on the inside of the test tube which is flushed with inert gas
b. Jars: the oxygen content of the jar is reduced by its conversion to water in the presence of introduced hydrogen using palladium as the catalyst and hastened by:
   — heat (McIntosh & Fildes 1916)
   — electricity (Brewer)
   — finely dividing the palladium over asbestos or pellets of alumina (Stokes 1958)
   Commercially available kits (GasPak®, GasKit®) generate hydrogen and CO_2 after the addition of water to dry packets of $NaHCO_3$ + $NaBH_4$. Widely used in clinical laboratories.
c. Chambers: cabinets and tents with entry locks and arm ports (glove boxes) are flushed with inert gases, providing a large enclosed space for culture manipulation in the absence of oxygen.

Indicators are used with all of these techniques to verify that anaerobic conditions have been provided, viz:
— methylene blue (blue — colourless at Eh 11 mv)
— reasurin (yellow — colourless at Eh − 51 mv)
— absence of growth of a strict aerobe (e.g. *Pseudomonas aeruginosa*)
— growth of a strict anaerobe (e.g. *Clostridium novyi*)

Some characteristics common to most strict anaerobes

1. Usually ferment carbohydrates (the major energy source) producing distinctive metabolic products used in classification.
2. Catalase is not detected (with the exception of *Propionibacterium*, *Actinomyces* and some *Bacteroides*).
3. Growth is usually poor in unsupplemented medium. Various special requirements including haeme, vitamin K, CO_2.
4. Molecular oxygen, superoxides and organic peroxides appear to inhibit or damage anaerobes. Strict anaerobes lack superoxide dismutase (which converts superoxide radicals to water and oxygen).
5. Anaerobes are usually resistant to aminoglycosides (kanamycin, neomycin, gentamicin), nalidixic acid, polymyxin and colistin. These antibiotics are therefore used in selective media.
6. Anaerobes are sensitive to metronidazole (with the exception of *Propionibacterium* and *Actinomyces*).
7. Growth is often enhanced in the presence of aerobic bacteria which scavage free oxygen and provide nutrients.
8. Pathogenic anaerobes are relatively aerotolerant (i.e. grow in 2–8% oxygen and tolerate exposure to air for 1–2 hours).

Anaerobes are classified initially using the same methods as those used for aerobes:

a. Gram reaction
b. shape
c. spore formation

and further classified using:

d. fermentation patterns of carbohydrates
e. types of compounds produced by the fermentation of glucose, using gas-liquid chromatography
   — volatile fatty acids (acetic, propionic, isobutyric, butyric, isovaleric, isocaproic, caproic)
   — non-volatile fatty acids (pyruvic, lactic, succinic)
f. ability to hydrolyse various proteins (milk, meat, serum, gelatin)
g. colonial morphology, pigment production
h. haemolysis
i. production of lecithinase and lipase on egg yolk agar
j. tolerance of bile
k. indole, urease production
l. hydrolysis of aesculin
m. reduction of nitrate
n. animal pathogenicity

Table 33 Classification

Gram-positive rods with spores	Clostridium	Produce exotoxins Resist heat
Gram-positive rods without spores	Actinomyces Arachnia Bifidobacterium Eubacterium	Slender with branching or bifid ends
	Propionibacterium	Club-shaped, Chinese characters, catalase positive
	Lactobacillus	
Gram-negative rods	Bacteroides Fusobacterium Leptotrichia Selenomonas	
Gram-positive cocci	Peptococcus Peptostreptococcus	
Gram-negative cocci	Veillonella	

Clostridia
— Gram-positive, generally large rods. May be Gram-negative in old cultures. Large spreading colonies
— Form spores (autoclaving is necessary for destruction)
— More than 60 species, some poorly defined
— Normal habitat is faeces and soil
— Not particularly invasive but cause disease by toxin formation, as follows:
1. Enterocolitis
   a. *Cl. perfringens (welchii)*
   b. *Cl. difficile*
2. Gas gangrene
   — follows multiplication of contaminating clostridia in devitalised tissues, especially muscle
   — circulating toxins cause haemolysis, septicaemia
   — usual causes are *Cl. perfringens, septicum, novyi, bifermentans*
3. Neurotoxic
   — two important syndromes (Table 34)

Table 34

	Tetanus	Botulism
Organism	Cl. tetani	Cl. botulinum
Origin	Animal faeces Soil	Soil
Pathogenesis	Contamination of wound, limited multiplication	Multiplication in preserved, especially alkaline, food
	Toxin enters circulation	Absorbed after ingestion
Toxin	Tetanospasmin, one antigenic type	Polypeptide of mol wt 150 000 7 antigenic types
Mode of action	Interferes with synaptic antagonists in spinal cord	Prevents release of acetylcholine at peripheral synapses
Incubation period	Days–weeks	12–36 h after food
Clinical features	Pain and rigidity of muscles, especially of the jaw. Later, reflex spasms	Diploplia, bulbar palsies. Later, paralysis of extremities
Vaccines	Hyperimmune globulin Toxoid	Hyperimmune globulin

Bacteroides species
— Frequently encountered Gram-negative rods
— Peritrichous flagella (although mostly nonmotile)
— Classified into three groups as follows:
  1. Fragilis group
     Ferment several sugars
     Bile tolerant, hydrolyse aesculin
     Produce beta Lactamases
     Predominant anaerobes of the large bowel
     Bact. fragilis is the most pathogenic
  2. Melaninogenicus group
     Ferment some sugars
     Form pigments, often fluorescent
     Dominant anaerobes of the oral cavity
  3. Others
     Bact. bivius, disiens, oralis, ruminicola, corrodens

Characteristics of infections due to non-sporing anaerobes

1. Infection occurs in relationship to mucosal surfaces where anaerobes are normal flora, viz:
   a. Oral cavity
      Periodontitis
      Aspiration pneumonia, lung abscess, empyema
      Brain abscess
      Ludwig's angina
   b. Colon
      Peritonitis, paracolic collections,
      Subphrenic abscess
      Surgical wound infection
   c. Female genital tract
      Puerperal fever
      Post-operative infection
   d. Skin
      Necrotising fasciitis
2. Haematoma formation is a frequent prelude to surgical infection
3. Tissue necrosis and abscess formation occur early
4. Gas may be present in involved tissues
5. Discharges and pus are often foul-smelling
6. Multiple anaerobic species are usually isolated
7. Infection does not respond to treatment with aminoglycosides

Antimicrobial agents for anaerobes

Benzylpenicillin
— is active against most anaerobes except *Bact. fragilis.*
— is usually effective for infections related to the oral cavity
Metronidazole
— is very active against all anaerobes except some Gram-positive cocci
— has no activity against aerobes
Clindamycin
— is active against most anaerobes. Transferable resistance has been detected in *Bact. fragilis.*
Cefoxitin, latamoxef
— resist hydrolysis by betalactamase of *Bact. fragilis* although MICs may be relatively high
— are of proven value in peritonitis
Chloramphenicol
— active against all anaerobes
— resistance does not develop
— widespread use is precluded by risk of marrow toxicity
Erythromycin
— about 50% of strains of *Bact. fragilis* may be resistant
Tetracycline
— is effective, but resistance is widespread

Mycobacteria

Gram-positive rods with complex cell wall structure which renders these organisms difficult to stain but which retains dyes strongly despite treatment with acid or acid and alcohol (i.e. they are acid-fast)

The mycobacterial cell wall
— consists of peptidoglycan covered by layers of rope-like peptido-glycolipids embedded in sulpholipids and trehalose-dimycolates
— contains abundant lipid, including genus-specific alpha hydroxy branched fatty acids of high molecular weight (mycolic acids)
— is relatively resistant to acids, alkalies, quaternary ammonium compounds, chlorhexidine, dyes, halogens and heavy metals
— has adjuvant properties (i.e. increases the antigenicity of foreign macromolecules)

Technique of the acid-fast stain
1. Aniline dye plus phenol at high concentration or with heating
2. Decolourise with acid or acid–alcohol
3. Stain the background with a dye of contrasting colour (counter-stain)

Precautions in the use of acid-fast stains
1. Environmental mycobacteria in water, blotting paper or staining solutions may contaminate the preparation
2. Mycobacteria may be transferred between slides if they are stained in the same container
3. Other acid-fast objects include spores, *Nocardia* species, sperm, protozoan cysts, hydatid hooklets and vegetable matter

Commonly used staining methods
Ziehl-Neelsen
— uses 0.3% carbol fuchsin, followed by heating for 3–5 minutes. Decolourized with 20% sulphuric acid (5% for *M. leprae*), counterstained with methylene blue or malachite green.

Auramine-rhodamine
— Dyes which make acid fast organisms easy to see at low
magnifications using fluorescence microscopy.
— 'Counterstained' with potassium permanganate to render tissue
and debris non-fluorescent
Kinyoun
— Uses strong (4%) carbol fuchsin without heating.

Growth requirements
Most species will not grow on simple media. Growth requirements
include the following:
— Carbon: glucose, glycerol or pyruvate
— Nitrogen: asparagine, glutamate, pyruvate
— Strictly aerobic but hastened by 6–8% CO_2
Selective agents are used to remove oral flora. They incude:
a. agents used to liquify sputum:
acids, alkalies, detergents
b. compounds added to media:
malachite green, naladixic acid, lincomycin
Frequently used media include:
Lowenstein–Jensen
— consists of whole eggs, asparagine, potato flour and glycerol
solidified by heat (inspissation)
— growth takes 2–6 weeks
Middlebrook's
— 3 types of clear, agar-containing media which facilitate early
detection of growth
Dubos
— liquid medium or agar-containing medium
Kirchner's
— liquid medium or agar-containing medium

Methods of identification (see Table 35)
1. Colonial morphology: *M. tuberculosis* is dry, wrinkled, buff-
coloured, friable, tenacious and granular
2. Temperature requirements for growth
— *M.avium* may grow poorly at temperatures of 37°C or lower
— *M. marinum* grows best at 24–31°C
3. Pigmentation of colonies
— produced in the dark (photochromogens)
— produced in the light (scotochromogens)
4. Rate of growth
— *M. fortuitum* and *M. chelonei* grow rapidly (3–5 days)
5. Production of niacin by *M. tuberculosis*
6. Aryl sulphatase test distinguishes 'rapid growers'
7. Catalase production
8. Growth on MacConkey' medium — by *M. chelonei* and *M.
fortuitum*

Table 35 Identification of the pathogenic mycobacteria

Species	Colonial morphlogy	Temperature range or optimum	Growth rate	Pigmentation	Niacin production	Aryl sulphatase	Catalase production	Growth on MacConkey
tuberculosis	Rough	37	Slow	None	+	−	−	−
bovis	Rough	37	Slow	None	−	−	−	−
ulcerans	Rough	37	Slow	None	−	−	−	−
leprae	Will not grow on laboratory media. Can be propogated in mouse food pad or armadillo.							
kansasii	Smooth	24–37	Slow	Photochromogen	+	−	+	−
marinum	Smooth	24–31	Moderate	Photochromogen	+/−	−	−	−
simiae	Smooth	37	Slow	Photochrogen	+	−	+	−
scrofulaceum	Smooth	24–37	Slow	Scotochrogen	−	−	+	−
szulgae	Smooth/Rough	24–37	Slow	Scotochromogen	−	−	+	−
xenopi	Smooth	37–45	Slow	Scotochromogen	−	−	−	−
avium-intracellulare	Smooth/Rough	37–42	Slow	None	−	−	−	−
haemophilum	Rough	24–31	Slow	None	−	−	−	−
fortuitum	Smooth/Rough	24–37	Rapid	None	−	+	+	+
chelonei	Smooth/Rough	24–37	Rapid	None	+/−	+	+	+

Table 36 Major groups of pathogenic mycobacteria

M. tuberculosis, M. bovis	Tuberculosis, a disease which can involve any tissue, e.g. lung, lymph nodes, meninges, bone, joints, urogenital, etc
M. kansasii, M. intracellulare, M. avium, M. chelonei, M. fortuitium, M. scrofulaceum, M. xenopii, M. simiae, M. szulgai	These mycobacteria are frequently isolated from environmental sources (e.g. water supplies) but sometimes cause disease (pulmonary, glandular, abscesses) especially (but not necessarily) in debilitated patients.
M. leprae	Leprosy, a generalised disease involving many tissues especially skin and peripheral nerves
M. marinum	Skin disease (swimming pool granuloma)
M. ulcerans	Tropical skin disease (Buruli ulcer)

Enviromental AFB, e.g. *M. phlei, M. smegmatis, M. gordonae*. When isolated from sputum, urine, gastric washings must be differentiated from pathogens listed above.

TUBERCULOSIS

Infection due to *M. tuberculosis* or *M. bovis*, most commonly involving the lung.

Pathogenesis of pulmonary tuberculosis
1. Persons with 'open' pulmonary tuberculosis produce, by their coughing, an aerosol containing tubercle bacilli.
2. Water droplets of less than 10 µm carring tubercle bacilli, when inhaled by persons in close proximity, lodge in the alveolar ducts and alveoli in the periphery of the mid and lower lung fields.
3. Bacilli are ingested by phagocytes but not destroyed.
4. Continued intracellular multiplication and development of delayed hypersensitivity creates (over a period of 4–6 weeks) an inflammatory response characterised by:
   — granuloma formation (giant cells, epithelioid cells and lymphocytes)
   — central necrosis (caseation)
   This small peripheral pulmonary granuloma is known as the Gohn lesion and may be visible on chest X-ray.
   During this period the tuberculin test becomes positive.
5. During the first 2–10 weeks, local lymph nodes are involved (this together with the peripheral primary focus is knows as the Ranke complex).
6. Bacilli are disseminated via the blood stream to other sites where multiplication may cause metastatic infection to develop
7. Both the primary and secondary sites of infection usually undergo resolution, fibrosis or calcification. Some organisms remain viable in the lesions.

8. These lesions reactivate months or years after the primary infection in 5–15% of subjects, most frequently in the metastatically seeded pulmonary apices (*not* the site of the primary lung lesion). Reactivation occurs at the rate of about 4% per annum in the first 2 years.

Other portals of entry
1. Pharynx   causing cervical adenitis
2. Small bowel causing tuberculous ileitis
3. Skin   causing lupus vulgaris

Factors leading to reactivation include:
a. Old age
b. Male sex
c. Race (Negroid or Asian) — but here re-infection may also be involved
d. Immunosuppression
e. Alcoholism
f. Pulmonary silicosis
g. Poorly controlled diabetes
h. Gastrectomy

Clinical features of tuberculosis in various organs

1. PULMONARY

Manifestations of lung disease may represent primary or reactivation disease.
a. Primary
   — Lower zone infiltrate possibly accompanied by hilar lymphadenopathy
   — Small peripheral caseous foci may rupture into the pleural space eliciting a copious exudate in which acid fast bacilli can not usually be found
   — Erythema nodosum accompanies primary infection in about 15% of cases.
b. Reactivation
   — Involves mostly the apical or posterior segments of the upper lobes because PO_2 is higher at these sites
   — Radiological appearance is of infiltrate or cavity later followed by contraction and distortion
   — Constitutional symptoms include fever, night sweats, malaise, weight loss
   — Pulmonary symptoms include cough, sputum production, haemoptysis, occasionally pleurisy
   — Spread of disease occasionally causes bronchial ulceration, lobar consolidation, empyema, laryngitis

2. Extrapulmonary
— Accounts for about 15% of all tuberculosis
— In developed countries, occurs in an older age group
— The tuberculin test is nearly always positive
— Incidence of constitutional symptoms depends on the organ involved. ESR is usually elevated
— Biopsy is usually required to establish the diagnosis

   a. Lymphadenitis
     — hilar and mediastinal
     — cervical

   b. Pyelonephritis
     — results in areas of caseation and calcification with later involvement of renal pelvis and ureter
     — insidious onset of dysuria, frequency, haematuria
     — investigation reveals proteinuria, sterile pyuria, renal calcification (16–50%), caliectasis, cortical scarring, ureteric strictures and cystitis
     — diagnosed by collection of at least three early morning samples of urine. Commensal mycobacteria of the lower urinary tract may be confused with *M.tuberculosis.*

   c. Pericarditis
     — follows spread from mediastinal lymph nodes or adjacent segments of lung to the pericardium
     — causes slowly developing tamponade, later constriction

   d. Epididymitis
     — results in a painful palpable epididymal mass
     — may coexist with renal tuberculosis

   e. Salpingitis
     — Fallopian tubes are most commonly involved with some extension into the uterus
     — causes pelvic pain, menstrual disorders and infertility
     — Granulomas may be detected in endometrial curettings obtained before the onset of menstruation in about 50%, otherwise laparotomy and tubal biopsy is required

   f. Arthritis
     — *M. tuberculosis* has a predeliction for the spine (Pott's disease) and weight bearing joints (hips, knees, ankles)
     — spinal disease causes narrowing of the interspace, erosion and destruction of adjacent vertebrae, scoliosis (gibbus) formation and paravertebral abscess
     — biopsy and culture establish the diagnosis

   g. Peritonitis
     — causes scattered peritoneal tubercles, omental thickening and exudate
     — resulting in abdominal pain and tenderness, distension, palpable masses, ascites
     — acid fast bacilli are not easily demonstrated in ascitic fluid. Biopsy at laparoscopy or laparotomy is required

h. Bowel infection
   — is usually associated with advanced pulmonary disease
   (i) ileocaecal: with circular mucosal ulceration, fibrosis, mesenteric lymphadenitis
   (ii) anorectal: with ulceration and stricture formation
i. Meningitis
   — usually a slow onset over several days with worsening fever and meningism
   — followed by cranial nerve palsies, raised intracranial pressure and obtundation
   — CSF examination reveals 100–500 cells/μl, mostly monocytic, high protein, low glucose
   — acid fast bacilli often cannot be demonstrated in CSF. Examination of the centrifuged deposit of large volumes increases the yield but treatment must often be commenced on clinical grounds
j. Intracranial tuberculoma
k. Miliary tuberculosis
   — Is disseminated disease with tubercles about the size of millet seeds (hence miliary) scattered throughout many organs
   — Usually has an insidious onset with fever, malaise, weight loss and no diagnostic features
   — usually occurs in patients with no history of prior tuberculosis
   — may cause hepatosplenomegaly, lymphadenopathy, meningism, choroidal tubercles and occasionally skin lesions
   — is associated with a negative tuberculin test in about 50%
   — has a well described but uncommon association with leukoerythroblastic anaemia
   — causes a miliary but often transient infiltrate on chest X-ray
   — is difficult to diagnose. Relevant sites of involvement should be cultured (sputum, urine, CSF, lymph nodes). Blind biopsy of bone marrow and especially the liver will usually reveal granulomas
   — may have to be treated empirically if there is sufficient clinical evidence to suggest the diagnosis.

Examination of sputum in suspected pulmonary tuberculosis:
1. Mucoid secretions are liquefied by NaOH or other liquifying agents.
2. If the method used for digesting sputum does not destroy oral flora, the specimen must be decontaminated.
3. The liquefied secretions are concentrated by centrifugation.
4. The concentrated deposit is streaked on Lowenstein–Jensen medium and incubated at 37°C for 4–6 weeks.

The tuberculin test
— utilises a Tween-stabilised purified protein derivative (PPD)
    which when injected intradermally detects delayed
    hypersensitivity
— causes a positive reaction 4–6 weeks after exposure and
    regardless of whether clinical disease occurs
— detects hypersensitivity which persists lifelong
— may detect cross reaction in subjects exposed to other
    mycobacteria (e.g. *M. avium-intracellulare*)

Methods for performing the test include:
1. Mantoux:
   Intradermal injection of 5 tuberculin units using a small volume
   (tuberculin) syringe. 10 mm of induration after 48–72 hours is
   regarded as positive
   The most sensitive test
2. Heaf:
   Uses a spring-loaded tuberculin-coated multipronged needle gun
   More convenient, less sensitive
3. Tine:
   Uses tuberculin-coated needle. Less sensitive

False negative tests may be due to:
1. Inadvertant subcutaneous administration
2. Use of poor quality tuberculin
3. Concurrent viral infection suppressing delayed hypersensitivity
4. Immunosuppression
5. Overwhelming disease, especially miliary tuberculosis

Repeated skin testing may stimulate an anamnestic ('booster')
response in previously infected subjects with initially weak
reactions.

Principles of management of tuberculosis
1. In order to prevent the development of resistance, the patient
   must be treated with at least two drugs to which the organism is
   sensitive. This usually means using three drugs until the results
   of sensitivity tests are known
2. Duration of treatment must be adequate
3. Compliance with therapy must be assured
4. By aleviating cough and reducing bacterial numbers in sputum,
   therapy renders the patient non infectious in about two weeks.
   Until then, infection control measures must be instituted
5. All contacts of the patient must be screened for acquisition of the
   disease (tuberculin test and chest X-ray)

Common treatment regimens (see Table 37)
1. Isoniazid 400 mg ⎫ daily for 9 months. The ethambutol is
   Rifampicin 450 mg ⎬ ceased if the organism is senstive to
   Ethambutol 1500 mg ⎭ isoniazid and rifampicin

2. Isoniazid ⎫ used when primary drug resistance is
   Rifampicin ⎬ suspected
   Ethambutol ⎪
   Pyrazinamide ⎭

3. Isoniazid ⎫ used in countries where the cost of
   Streptomycin ⎬ rifampicin precludes its use
   Thiacetazone ⎭

4. Regimens of intermittent therapy are used when administration
   must be supervised, ie, the patient is watched while the tablets
   are swallowed. All are given twice weekly.
   a. Streptomycin 1 g i.m.   plus   isoniazid 15 mg/kg   18 months
   b. Ethambutol 50 mg/kg   plus   isoniazid 15 mg/kg   18 months
   c. Pyrazinamide 50 mg/kg   plus   isoniazid 15 mg/kg   18 months
   d. Rifampicin 600–900 mg   plus   isoniazid 15 mg/kg   12 months

5. Recent work has shown that short-course chemotherapy using
   four drugs is effective in respiratory tuberculosis; isoniazid,
   rifampicin, pyrazinamide and streptomycin for two months
   followed by isoniazid and rifampicin for a further 4–6 months

Table 37 Antimicrobial agents for mycobacterial infections

Drug	Mode of action	Dose and route	Side effects
Isoniazid	Cidal	300–400 mg p.o. daily	Hepatitis 0.3 % Hypersensitivity Neuropathy (prevented by pyridoxine) Haemolytic anaemia
Rifampicin	Cidal	450–600 mg p.o. daily	Hepatitis in 3%, more common in aged and alcoholic Hypersensitivity Gastrointestinal upset Antagonises effect of oral contraceptives
		600–900 mg twice weekly	Flu-like febrile illness Pancreatitis Hepatitis Thrombocytopenia Acute renal failure Asthma Pruritis and rash

Table 37 Contd

Drug	Mode of action	Dose and route	Side effects
Ethambutol	Static	25 mg/kg/day for 2 months p.o., then 15 mg/kg/day	Optic neuritis Hypersensitivity
Streptomycin	Cidal	0.75–1.00 g i.m. daily	Vestibular damage especially in old age or in the presence of renal failure
Pyrazinamide	Cidal	1.5–2.0 g p.o. daily	Hepatitis Hyperuricaemia Hypersensitivity Renal damage
Prothionamide Ethionamide	Static	500–750 mg p.o. daily	Gastrointestinal upset Hepatitis Metallic taste Headaches
Cycloserine	Static	0.5–1.0 g p.o. daily	Seizures (may need anti-convulsants) Depression, psychosis
Capreomycin Kanamycin Viomycin	Cidal	1 g i.m. daily	Auditory damage Occasional renal damage
Thiacetazone	Cidal	150 mg p.o. daily	Hepatitis Haemolytic anaemia
Para-amino salicylic acid	Static	12 g p.o. daily	Gastrointestinal upset
Dapsone	Static	50–100 mg p.o. daily	Haemolysis Methaemaglobinaemia Gastrointestinal upset Hypersensitivity
Clofazamine	Static	100–300 mg p.o. daily	Skin pigmentation

LEPROSY

— Caused by *M. leprae* (Hansen's bacillus) which can only be propogated in armadillos and the footpads of mice
— Less acid-fast than *M. tuberculosis* (the Ziehl–Neelsen stain should be decolorised in 1% rather than 3% acid–alcohol)
— Divides slowly (13 days)
— The epidemiology is not well understood but it is likely that the profusely colonised nasal secretion of patients with lepromatous leprosy is the source of most infections and that the portal of entry is the respiratory tract

— Multiplies in the cooler parts of the body, mostly the skin, with a predeliction for peripheral nerves
— Clinical and pathological manifestations are the consequence of immune status. Five groups have been defined (Ridley & Jopling 1966)

1. Full tuberculoid (TT)   Most limited lesions
2. Borderline tuberculoid (BT)   May change from one type to
3. Borderline (BB)   another during the course of
4. Borderline lepromatous (BL)   treatment
5. Full lepromatous (LL)   Most generalised lesions
   The basis for the classification is given in Table 38.

Clinical features
1. Skin lesions:
   Hypopigmented macules, plaques, nodules
2. Nerve deficits (motor and sensory)
3. Lesions due to hypersensitivity ('lepra reactions')
   Erythema nodosum

Diagnosis
1. Skin biopsy (and acid fast stain)
In *tuberculoid* leprosy, biopsy of skin lesions reveals changes very similar to sarcoidosis but with nerve bundles grossly swollen and infiltrated with mononuclear inflammatory cells. Acid fast bacilli are few or absent.
   In *lepromatous* leprosy, biopsy reveals granulomatous dermal infiltrates consisting of foamy macrophages laden with acid fast bacilli.

2. Skin scrapings
In lepromatous leprosy, smears made from scrapings of dermis from ear lobes, nares and eyebrows reveal numerous acid fast bacilli.

3. Nerve biopsy
In tuberculoid leprosy, biopsy of nerves demonstrated to have abnormal conduction times may reveal typical changes.
   The *Morphological Index* is calculated by noting the proportion of uniformly staining (live) bacilli compared to the number of irregularly staining (dead) bacilli in skin biopsies.

   The *skin sensitivity* (lepromin) test helps to classify the type of disease but is not useful diagnostically.

Table 38 Classification of leprosy

	Tuberculoid	Lepromatous
Skin lesions	Discrete and few	Diffuse but less disfiguring
Histology	Sarcoid-like with lymphocytes and giant cells	Foamy macrophages
No of bacilli	Sparse, usually not detectable	Profuse
Nerve involvement	Confined to one or two, greatly thickened.	Diffuse
Systemic disease	Nil	Granulomatous lesions in liver, spleen and lymph nodes.
Delayed hypersens to antigens of M. leprae	Present	Absent (anergic)
Immunoglobulin levels	Normal	Polyclonal hyper gammaglobulinaemia
Autoantibodies	Absent	Present
Amyloid-related protein	Absent	Present
Circulating immune compexes	Absent	Present
Complications	Related mainly to nerve involvement	Erythema nodosum Uveitis Amyloid
Infectivity	Low	High (?via nasal discharges)
Racial distribution		
Indians	80%	20%
Africans	90	10
Whites and Orientals	50–70	30–50

Complications
1. Erythema nodosum leprosum
2. Deformity
   — of nose
   — of hands and feet
3. Amyloidosis
4. Eye disease
   — conjunctivities
   — superficial punctate keratitis
   — iridocyclitis

Principles of management
1. To prevent the emergence of drug resistance, all patients with lepromatous (i.e. multibacillary) disease should receive three drugs for an adequate period of time.
2. In those few centres which can afford it, sensitivity tests using the mouse foot pad should be undertaken on new cases. Dapsone resistance is now a major problem.
3. The adequacy of treatment of lepromatous disease should be followed by examining the Morphological Index on specimens collected during the course of treatment. In untreated cases the value is between 25 and 75 but falls to zero after 6 months of effective therapy.
4. Immune complex disease should be anticipated during the treatment of lepromatous disease and treated, if necessary, with steroids or thalidomide.
5. Therapy renders the patient non infections. Although patients are better managed in centres with frequent experience of the disease, incarceration and segregation from the family or community is undersirable.
6. Deformity and disability should be minimised as follows:
   a. protection of anaesthetic areas
   b. patient education
   c. surgery (eg, nerve transplants)

Regimens for chemotherapy

Tuberculoid	Dapsone	1–2 mg/kg daily for 6 months
	Rifampicin	600 mg monthly for 6 doses
Lepromatous	Dapsone	100 mg daily ⎫ for a minimum of 2
	Rifampicin	600 mg monthly ⎬ years and until
	Clofazamine	300 mg monthly ⎭ smear-negative

Ethionamide or prothionamide 250–375 mg daily can replace clofazamine.

Spirochaetes, myco plasmas, chlamydiaceae, rickettsieae

SPIROCHAETES

A large heterogenous group of spiral-shaped organisms with a central contractile axostyle which creates rotating and flexuous movements.

Cell wall is Gram-negative in type, but spirochaetes with a transverse diameter near the resolution of the light microscope are best observed by dark ground illumination or silver stains. The genera of medical interest are given in Table 39.

Table 39

Genus	Morphology	Habitat	Diseases
Treponema	Non-cultivatable, highly coiled, 10–20 μm	Pathogenic and non-pathogenic species of man	Syphilis Yaws Pinta
Leptospira	Cultivatable, 5–20 μm, hooked ends	Two species Zoonotic	Leptospirosis
Borrelia	Loosely coiled, 8–16 μm, stain with aniline dyes	Transmitted from man to man by lice and ticks	Relapsing fever Lyme disease
Intestinal spirochaetes	Two or three short curves, 3 μm	? Normal inhabitant of bowel	? Proctocolitis

Treponematoses
Caused by morphologically similar, non-cultivatable spirochaetes which share antigens and produce serological responses which are cross-protective and cannot distinguish one disease from another. Three species have been defined on the basis of host range and clinical picture of the diseases produced (Table 40).

Diagnosis of treponematoses
1. Dark ground illumination of exudate from lesions
2. Serology

74

Table 40

Species	Disease	Other hosts	Distribution
T. pallidum	Syphilis	Rabbits	Worldwide, venereal
	Bejel		Endemic syphilis is transmitted amongst children in arid parts of Africa and Arabian peninsula
T. pertenue	Yaws	Rabbits Hamsters	Throughout the tropics
T. carateum	Pinta	Nil	South America

Two types of antigen are used. Neither can distinguish between infections due to the three *Treponema* species.

a Reaginic
- depends upon cross-reactivity between *Treponema* and an antigen (cardiolipin) derived from beef heart muscle
- starts to become positive with the appearance of the chancre but not 100% seropositive until 4 weeks
- may be negative in about one third of patients with late syphilis
- are measured in titre to follow the progress of treated syphilis
- includes several techniques in which cardiolipin is combined with lecithin, cholesterol and a carrier, viz.,

VDRL	Venereal Diseases Research Laboratory	Flocculation on glass slide
RPR	Rapid Plasma Reagin	Flocculation. Charcoal carrier makes the test easy to read
WR	Wassermann reaction	CFT introduced in 1906, now superseded by above tests

Kline ⎱
Kahn ⎬ Previously used tests, now only of historical significance
Kolmer ⎰

- are non-specific tests. False positive reactions occur in up to 10% of specimens, due to:
  (i) Viral infections, especially infectious mononucleosis, measles, varicella
  (ii) Vaccination
  (iii) Some bacterial infections, especially leprosy, typhus, lymphogranuloma venereum
  (iv) Autoimmune diseases, such as systemic lupus erythematosis
  (v) Pregnancy

b. Specific antitreponemal tests
   — utilise Nichol strain of *T. pallidum* propogated in rabbit testis
   — have a much lower false positive rate (about 1–2%) than
     reaginic test
   — include the following:

FTA–ABS	Fluorescent treponemal antibody absorption test	Patient's serum is first absorbed with nonpathogenic treponemal antigen. Slides containing smears of *T. pallidum* are used in IFA test
TPHA	*T. pallidum* haemagglutination assay	Serum is treated as above then utilised in PHA test
TPI	*T. pallidum* immobilisation test	Presence of antibody arrests freshly collected motile spirochaetes

Genus Leptospira
   Cultivatable spirochaetes with hooked ends. Strictly aerobic but
grow best in liquid media supplemented with serum.
   The genus has two species:
L. biflexa: free-living saprophyte
L. interrogans: pathogenic for man and animals. 135 serotypes
which were previously designated as species.

Leptospirosis
Transmitted to man from animals (rodents, cats, dogs, pigs, cows)
which excrete the organism in urine for many weeks after infection.
   Penetrates the skin, causes a flu-like illness which may progress to:
1. Hepatitis
2. Renal failure
3. Meningitis
4. Lymphadenopathy
   Diagnosed by:
1. Dark ground examination of plasma or urine early in the course of
   infection.
2. Blood culture (Korthoff's medium, guinea pigs)
3. Serology: development of type-specific agglutinins. Groups of
   antigens are chosen according to prevalent serotypes in a given
   region
   Treatment of choice is benzyl penicillin.

Borrelia species
Species of medical interest are shown in Table 41.

Table 41

Species	Vector	Disease	Habitat
B. recurrentis	Louse	Relapsing fever	Endemic in parts of Africa, S. America and S. E. Asia
B. duttoni	Tick	Relapsing fever	Foci throughout the world
Borrelia species (unnamed)	Tick	Lyme disease	? Worldwide
B. vincenti		Vincent's angina 'tropical ulcers'	Oral commensal

Relapsing fevers
— Incubation period of about 7 days followed by fever, myalgia, arthralgia, hepatosplenomegaly. Jaundice may follow
— Damage to capillary endothelium may cause petechiae, bleeding
— Remission follows antibody formation but about a week later following emergence of a new antigenic variant, relapse occurs. The cycle may be repeated several times
— Diagnosis is made by finding spirochaetes in stained films of blood
— Treatment with penicillin, tetracycline or chloramphenicol is effective but is accompanied by a Jarisch–Herxheimer reaction due to accelerated release of endotoxin

Lyme disease
An infection first recognised in Lyme, Connecticut in 1975. Characterised by:
1. Skin lesion (erythema chronicum migrans). An annular lesion developing at the site of the tick bite
2. Constitutional symptoms (fever, headache, myalgia)
3. In some patients, especially with B-cell alloantigen, DR2, visceral complications may occur months to weeks later. viz.,
   a. Arthritis — intermittent, involves large joints
   b. Meningoencephalitis, peripheral neuritis
   c. Myocarditis
Diagnosis is made by indirect IFA using the Lyme spirochaete.
Responds to penicillin, tetracycline.

MYCOPLASMAS

— Small bacteria lacking a cell wall and therefore resistant to beta lactam antibiotics
— Grow on supplemented media with a tendency to grow down into the agar producing colonies with a dark centre and light periphery ('fried egg' appearance)
— Fermentative, facultative. All require sterols for growth
— Specific antibody prevents growth
— Human mycoplasmas include two genera (Table 42)

Table 42

	Colonies	Substrates	pH opt.	
Mycoplasma species	large	glucose arginine	7.0	
Ureaplasma urealyticum	tiny	urea	6.0	Previously called T-strains (T = tiny)

The clinical significance of the human mycoplasmas is given in Table 43.

Table 43

	Habitat	Disease
M. pneumoniae	Throat	Pharyngitis, pneumonia
M. salivarium M. orale	Throat	Commensals
M. hominis U. urealyticum	Genital tract	? Non-gonococcal urethritis ? Salpingitis, endometritis, chorioamnionitis

Mycoplasma pneumonia
— Eaton's pneumonia, 'primary atypical pneumonia', is caused by *M. pneumoniae*
— is characterised by family outbreaks of upper and lower respiratory tract infection
— accounts for about 20% of community-acquired pneumonia
— has an incubation period (3 weeks) followed by the insidious onset of pharyngitis, bronchitis and pneumonia
— causes segmental, sometimes multilobar or diffuse, consolidation
— may rarely be complicated by the following:
  1. Haemolytic anaemia
  2. Meningoencephalitis Polyneuritis
  3. Stevens-Johnson syndrome
  4. Pericarditis, myocarditis

OBLIGATORILY INTRACELLULAR BACTERIA

— Coccoid bacteria with gram negative cell walls, incapable of synthesising ATP, multiply only in host cells
— May be cultivated in embryonated hens' eggs or vertebrate cell cultures

Table 44 Obligatorily intracellular bacteria

Genus	Cell preference	Diseases
Chlamydia	Macrophages	Ornithosis Urethritis, conjunctivitis Neonatal conjunctivitis, pneumonia
Rickettsia	Arthropods Capillary endothelium	Typhus and related infections
Coxiella	Various	Q fever
Bartonella	Red cells	Bartonellosis (Carrion's disease, Oroya fever)

Chlamydiae

— Multiply in phagosomes of phagocytic cells. Phagosome does not fuse with lysosome
— Form cytoplasmic inclusions detected by staining
— Developmental cycle:
  Elementary body: is the extracellular infectious particle, which attaches to and enters susceptible cells
  Initial body: is the intracellular form. Metabolically, active, undergoes binary fission, later forms, elementary bodies
— Consists of two species having a common cell wall antigen but clinical, cultural and epidemiological differences (Table 45).

Table 45

	C. psittaci	C. trachomatis
Usual habitat	Birds	Man
Staining of inclusions with iodine	No	Yes
Activity of sulphonamides	Resistant	Sensitive
Clinical syndromes	Pneumonic	Venereal Ocular

C. trachomatis is divided serologically into 3 groups, viz:
1. Types A,B,Ba,C
   Hyperendemic blinding trachoma
2. Types D,E,F,G,H,I,J,K
   Inclusion conjunctivitis and genital tract diseases.
3. Types L−1,L−2,L−3
   Lymphogranuloma venereum

Chlamydial syndromes due to serotypes D to K
Males   Nongonococcal urethritis
               Epididymimoorchitis
Females   Cervicitis
               Urethral syndrome
               Salpingitis
               Bartholinitis
Neonates   Inclusion conjunctivitis
               Pneumonia (follows inclusion conjunctivitis in 50% of
               cases)

RICKETTSIEAE

— Short rods, 600 × 300 nm, often pleomorphic.
— Gram negative cell walls which share some antigens with
  Proteus species. Stain well with Giemsa stain.
— Cultivatable only in cells such as embryonated eggs (yolk sac)
  and vertebrate tissue cultures.
— Consists of two genera with different cell wall structures and
  epidemiology (Table 46).

Table 46

	Rickettsia	*Coxiella*
Cell wall	Readily leak macromolecules. Unstable outside host cells	Resist drying
Mode of transmission	Arthropods	Air
Portal of entry	Skin (bite)	Lung Ingestion
Clinical features	Rash	Pneumonia

Clinical features of rickettsial infection (see Table 47):
1. Eschar—at the site of arthropod bite
2. Fever—with abrupt onset, photophobia
3. Rash—4–7 days after the onset of fever
Diagnosis is usually made serologically rather than by isolation of
the organism, using the following techniques:
1. Weil-Felix agglutination test
   — depends upon cross reactions between cell wall of certain
     strains of *Proteus* species and various rickettsiae
   — is useful as a screening test but cannot differentiate disease
     due to different species
2. Rickettsial antigens, type and group specific
   — uses CFT, PHA, IFA
Treatment is with tetracycline or chloramphenicol

Table 47 Rickettsial diseases of man

	Species	Disease	Distribution	Arthropod vector	Vertebrate host	Proteus antigen
Typhus group	R. prowazecki	Epidemic typhus Brill–Zinsser disease	Potentially worldwide	Body louse	Man	OX 19
	R. mooseri	Endemic typhus	Worldwide	Rat flea	Rat	OX 19
Spotted fever group	R. rickettsii	Rocky Mountain spotted fever	Western Hemisphere	Tick	Various	or
	R. sibirica	Siberian tick typhus	Russia	Tick		OX 2
	R. conori	Boutonneuse fever	Mediterranean	Tick		
	R. australis	Queensland tick typhus	Australia	Tick		
	R. akari	Rickettsial pox	USA, USSR	Mite	Mouse	Nil
Scrub typhus	R. tsutsugamushi	Scrub typhus	Oceania	Mite	Small rodents and birds	OX K
Trench fever	R. quintana	Trench fever	Europe, during wars	Body louse	Man	Nil
Q fever	C. burnetii	Q fever	Worldwide	Tick	Wild and domestic animals	Nil

Q ('QUERY') FEVER

— Caused by *C. burneti*, an infection maintained worldwide by tick-borne transmission of the organism between wild and domestic animals
— Infects man by inhalation of dried secretions or ingestion of the contaminated milk of domestic animals
— After an incubation period of 7–15 days, causes fever and pneumonia
— May colonise damaged cardiac valves, causing a culture-negative endocarditis probably synonymous with 'chronic' disease
— Is diagnosed by:
   1. Isolation of organism but requires use of guinea pigs or embryonated eggs and may be transmitted to laboratory staff.
   2. Complement fixation test, using one of two antigens, viz.
   Phase 2
      Prepared from strains passaged in the lab
      Reacts with sera from acute and chronic disease
   Phase 1
      Prepared from freshly isolated strains in which the outer (phase 1 antigen) is preserved
      Reacts only with sera of patients with chronic disease
   Treatment is with tetracycline, chloramphenicol or cotrimoxazole.

Viruses — general characteristic

'Viruses are entities whose genome is in an element of nucleic acid, either DNA or RNA, which reproduces inside living cells and uses synthetic machinery to direct the synthesis of specialised particles, the virions which contain the viral genome and transfer it to other cells (Luria 1967)

The structure of viruses
1. Nucleic acid–consists of either DNA or RNA
   DNA — is double-stranded (except for parvoviruses)
   RNA — is single stranded (except for reoviruses)
       — is sometimes segmental (e.g. influenza)
2. Capsid
   — is the protein coat surrounding the nucleic acid.
   — consists of units of protein (capsomeres) arranged
       symmetrically around the nucleic acid in one of two ways:
   Incosahedral (cubic): The capsomeres form polyhedrons with
       20 triangular faces and 12 corners

   Helical: The capsomeres are applied directly to the, helical
       nucleic acid
3. Envelope
   — a partially host-derived lipid membrane also including an
       inner layer of protein and protruding glycoprotein spikes
   — is always present on viruses with helical capsomere
       symmetry but not always on those with cubic symmetry
   — renders the virion sensitive to lipid solvents such as ether
Viruses are classified initially according to above characteristics (see Table 48 and 49)

Table 48 Classes and structure of the viruses of man

Class	Mol.wt of nucleic acid (mult 10^6 daltons)	Particle size (nm)	Structure	Genera
DNA VIRUSES (in order of size)				
Poxviruses	130–240	170–260	Complex structure of several layers. Contains lipids and enzymes. Contains lipids and enzymes, including RNA polymerase. Cytoplasmic multiplication	Variola Vaccinia Cowpox Orf Molluscum contagiosum
Herpesviruses	54–92	130	Icosahedron in a lipid envelope. 162 capsomeres. Nuclear multiplication	Herpes simplex Epstein-Barr virus Varicella-Zoster Cytomegalovirus
Adenoviruses	20–30	70–90	Icosahedron with 252 capsomeres. Nuclear multiplication	Adenoviruses
Hepadnaviruses	(3182 base pairs)	40–47	Small circular, partially double-stranded DNA. Contain DNA polymerase	Hepatitis B
Papovaviruses	3–5	45–55	72 capsomeres in a skew arrangement. Assembled in nucleus	Papillomavirus Polyomavirus
Parvoviruses	1.5–2.2	18–26	Single-stranded DNA. Icosahedron with 32 capsomeres. Nuclear multiplication	Serum parvovirus Enteric parvovirus (es)
RNA VIRUSES				
Paramyxoviruses	5–8	150	Lipid envelope with spikes. Nucleocapsid has transcriptase activity, some have neuraminidases	Measles Mumps Parainfluenza Respiratory syncytial
Orthomyxoviruses	4	100	6 segments of RNA which assort to form genetically stable hybrids in mixed infections. Lipid capsule has spikes and neuraminidase. Helical nucleocapsid has transcriptase activity. Multiplication requires the nucleus	Influenza A Influenza B Influenza C

Coronaviruses	9	70–120	Enveloped particles with large spikes	Coronavirus
Arenaviruses	5.5	50–300	5–7 segments of RNA, enveloped. Contain ribosomes and a transcriptase	Lymphocytic choriomeningitis Lassa fever Argentine H F Bolivian H F
Retroviruses	6–10	100	Icosahedral, enveloped. Have an RNA-dependent DNA polymerase. DNA provirus is nuclear	Human T cell leukaemia virus (HTLV) Lymphadenopathy-associated virus (LAV)
Reoviruses	10–16	60–80	Icosahedron. 10–12 segments of double-stranded RNA. Cubic nucleocapsid with transcriptase activity	Rotavirus Orbivirus Orthoreoviruses
Picornaviruses	2.5	20–30	Small icosahedron. Cytoplasmic multiplication	Echoviruses Poliovirus Coxsackievirus Rhinovirus Hepatitis A
Rhabdoviruses	3.5–4.6	130–300	Bullet-shaped particles with spikes. Helical nucleocapsid with transcriptase activity. Cytoplasmic multiplication	Rabies
Togaviruses	4	40–70	Enveloped icosahedron	Alphaviruses Flaviviruses Rubella
Bunyaviruses	6	100	3 segments of RNA, lipid envelope has spikes	Hantavirus Bunyavirus Nairovirus Phlebovirus Uukuvirus
Filoviruses		80	Helical capsid with cross striations surrounded by envelope with spikes. Branching and bizarre shapes	Marburg virus Ebola virus.

Table 49 Initial classification of viruses

	Cubic enveloped	Cubic naked	Helical enveloped	Complex
DNA	Herpesviruses	Adenoviruses Papovaviruses Parvoviruses		Poxviruses Hepadnavirus
RNA	Togaviruses Retroviruses	Reoviruses Picornaviruses	Paramyxoviruses Orthomyxoviruses Arenaviruseses Rhabdoviruses Bunyaviruses Filoviruses	

The nature of virus-cell interactions

1. Attachment
   — electrostatic forces bind the virion to receptors on the cell surface. Antibody inhibits this process
   — in the case of influenza, a viral surface protein, neuraminidase acts on cellular glycoprotein
2. Penetration
   — by a process of pinocytosis or phagocytosis, creating a cytoplasmic vacuole
3. Uncoating
   — of viral nucleic acid by removal of lipid membrane and protein capsid
4. Replication
   — describes the process by which viruses reproduce by making copies of themselves using the metabolic processes of the host cell rather than multiplying by binary fission like bacteria
   — utilises mechanisms which depend upon the virus structure
   a. DNA viruses
      — the viral DNA template is transcribed to form early messenger RNA (mRNA) catalysed by a host-derived DNA-dependent RNA polymerase
      — early mRNA codes mainly for enzymes used in viral DNA replication, including DNA-dependent DNA polymerase
      — molecules of newly formed progeny DNA are transcribed to form late mRNA which codes for the formation of viral capsid proteins, using host cell ribosomes
   b. RNA viruses employ two main methods of replication
      (i) Single-stranded viral RNA attaches to host cell ribosomes and, acting as its own template, is translated into virusspecific proteins including RNA-dependant RNA polymerase. RNA-dependant RNA

polymerase produces RNA molecules which act as a template for virus replication as well as mRNA for protein synthesis

(ii) Myxoviruses contain a RNA-dependant RNA polymerase within the virion which transcribes mRNA from the virion RNA

5. Assembly, maturation and release of virions
   — The nucleic acids of most DNA viruses except poxviruses are synthesised in the nucleus while the capsid proteins are manufactured in the cytoplasm
   — Viral envelopes are acquired as virions bud through cell membranes

Clinical patterns of viral infection

1. *Mucosal*
— short incubation period, do not invade.
— e.g. respiratory viruses: influenza
   causes of diarrhoea: rotaviruses, parvoviruses

2. *Systemic*
— virus is inhaled or ingested then multiplies in epithelium and local lymphoid tissue,
— followed by multiplication in regional lymph nodes and
— transient viraemia (causing minor, non-specific symptoms)
— resulting in the seeding of virus to target organs where replication may occur as follows:
Skin
   — exanthem, e.g. measles, chickenpox
Mucous membranes
   — enanthem, e.g. measles, chickenpox
Meninges
   — meningitis, e.g. enteroviruses
Brain
   — encephalitis, e.g. arboviruses
Placenta
   — fetal damage, e.g. rubella
Liver
   — hepatitis, e.g. hepatitis A and B
Incubation period of these infections is 6–160 days

3. *Recurrent, recrudescent*
— follows activation of virus at sequestered sites, e.g.
   Herpes simplex: Cold sores
   Varicella-zoster: Shingles
   Cytomegalovirus: Pneumonia (in immunosuprocced)
   Epstein–Barr virus: Lymphoproliferative disease
   (in immunosuppressed)

4. *Slow virus infections (see below)*

5. *Post-infectious neurological syndromes*
    — follow infection by about 10 days and are thought to be due to hypersensitivity
    — includes encephalomyelitis, polyneuropathy

6. *Neoplasia (see below)*

Diagnostic methods in virology

1. *Microscopy*
a. The virions of poxviruses are large enough to be seen with light microscopy (Guarnieri bodies) stained by Gispen's method
b. Inclusion bodies are cellular aggregates of viral material detected by light microscopy and seen in the:
    — nucleus (herpesviruses)
    — cytoplasm (rabiesvirus, poxviruses)
c. Electron microscopy
    — is useful for skin lesions due to large viruses with cubic symmetry which create aggregates of virions in crystalline array (e.g. herpesviruses)

2. *Culture requires living cells, as follows:*
a. embryonated hen's eggs
b. tissue culture
    3 cell types are commonly used:
     (i) transformed (malignant) cell lines
    (ii) human embryonic fibroblasts
    (iii) monkey kidney cells

    Detection of viral growth in tissue cultures is achieved by
     (i) cytopathic effect (CPE)
        — cell death, rounding, shrinkage, removal from glass
        — production of multinucleate giant cells (syncytia)
        — transformation
        — some viruses produce no cytopathic effect
    (ii) haemadsorption
        — depends upon the ability of many enveloped viruses to attach to red cells of various animal species
    (iii) interference
        — occurs when a cell which is supporting the replication of one virus will not support the replication of a second virus
c. organ culture
    — utilises whole tissues, such as tracheal mucosa, to support the replication of viruses
d. animals
    — may be necessary for arboviruses, Coxsackie A viruses

3. *Detection of viral antigens*
e.g.
Hepatitis B surface antigen: serum
Rotavirus antigen: faeces
Respiratory syncytial virus: nasopharyngeal epithelial cells
Smallpox antigen: vesicle fluid
Rabies virus: neurones
4. Detection of viral nucleic acid by nucleic acid hybridisation
   — is a recently developed technique which may eventually
      replace many existing techniques
5. Antibody response
   — appears about 10 days after exposure
   — in primary infections is dominated in the acute period by IgM
      class (provides a rapid test of infection on single early
      samples of serum)
   — can be assayed by several methods:

Complement fixation	Widely used, produces fairly short-lived high titres. Not suitable for IgM class
Haemagglutination-inhibition	Useful for viruses which haemagglutinate red cells. Non-specific serum inhibitors may interfere
ELISA, RIA	Are increasingly used. Can be readily used to detect IgM fraction
Immunofluorescence	Useful for detecting antibodies to viral antigens present in certain parts of infected cells (e.g. Epstein–Barr nuclear antigen)
Neutralisation	Time-consuming but highly specific

Haemagglutination by viruses
Many clases of viruses attach to red cells of various species,
creating, in the suspending fluid, an agglutinated mass of cells. The
process is inhibited by specific antiviral antibodies.
   Viruses which agglutinate red cells include:
1. Myxoviruses—via glycoprotein spikes on surface of virion
2. Togaviruses
3. Adenoviruses—via penton fibre
4. Papovaviruses

SLOW VIRUS DISEASES

Progressive pathological processes caused by transmissible agents which remain clinically silent during an incubation period exceeding many months or years.

Most involve the central nervous system, as follows:
1. Subacute sclerosing panencephalitis (SSPE)
   — Measles
   — Rubella
2. Rabies
   — Rabies virus
3. Progressive multifocal leucoencephalopathy (PML)
   — JC virus
4. Transmissible spongiform encephalopathies
   — Unconventional viruses

Unconventional viruses

The term was coined by Gadjusek to describe transmissible agents with the following characteristics:
1. Small size 25–100 nm
2. Unusual resistance to chemical and physical agents
3. Failure to induce inflammatory or immune response
4. Lack of demonstrable nucleic acids
5. Predeliction for brain involvement with titres of up to 10^8 infectious particles per gram of infected material
6. Pathology is characterised by diffuse loss of neurones, astrocyte proliferation and fibrous gliosis

Four diseases have been delineated:
1. Kuru
2. Creutzfeldt–Jakob disease
3. Scrapie (in sheep)
4. Transmissible mink encephalopathy (in mink)

The human diseases have the characteristics set out in Table 50.

Table 50

	Kuru	Creutzfeldt-Jacob disease
Geographical distribution	Fore tribe of the East New Guinea highlands	Worldwide with some small geographical clusters (e.g., Libyan Jews)
Age of onset	Children and adolescents	40–60
Sex	M:F = 1:3	M:F = 1:1

Table 50 Contd

	Kuru	Creutzfeldt-Jacob disease
Mode of transmission	Contamination of skin cuts and mucosa with the brain tissue of dead kinsmen during ritual cannibilism	Direct inoculation Corneal transplant
Secondary spread	No	10% are familial
Incidence	Disappearing in last 15 years	One per million of population
Experimental	Primates, incubation period of 18–30 months	Primates, cats, mice and guinea pigs
Clinical features	Cerebellar ataxia Tremor Death in 3–9 months	Dementia Ataxia Myoclonic jerking Death in less than 1 year

Viruses and malignancy
Despite numerous examples of viruses which causes malignancy in other vertebrates, it has been difficult to establish the same for man. The following viruses have been implicated as cause of malignancy but the evidence for some has been circumstantial
1. Epstein–Barr virus
   a. Burkitt's lymphoma
      — is a common cause of lymphoma in Africa in a zone corresponding to that of malaria
      — tumour tissue yields Epstein–Barr virus and patients have high circulating antibody titres to EBV.
   b. Nasopharyngeal carcinoma
      — is a common carcinoma in S.E. Asia
2. Hepatitis B virus
   — is associated with hepatoma in subjects with prolonged carrier state
3. Herpes simplex, type 2
   — has been linked with carcinoma of cervix
4. Papillomavirus
   — has been linked with carcinoma of cervix
5. Human T-cell leukaemia virus (see chapter on Retroviruses)

Chemotherapy of viral infections

Table 51 The classes of antiviral agents

Substance	Structure	Mode of action	Spectrum
Interferons	Proteins	Stimulate production of proteins which inhibit translation of viral m-RNA	Broad
Nucleoside analogues	Substituted derivatives of purines and pyrimidines	Inhibit nucleic acid synthesis	Mostly herpes viruses
Amantadine	3-ringed amine	Prevents uncoating of virus	Influenza
'Foscarnet'	trisodium phosphono-formate	Inhibits viral DNA replication	Herpesviruses

Nucleoside analogues
These are the only widely used antiviral agents to date.

Idoxuridine Trifluorothymidine	Applied as eye drops for the treatment of herpetic keratitis. Too toxic for systemic use
Vidarabine (Adenine arabinoside)	Used systemically it reduces mortality and morbidity of infections due to herpes simplex and varicella-zoster. Relatively insoluble. Neurotoxic and cytotoxic
Acyclovir (Acycloguanosine)	In infected cells is converted by herpes virus-specific thymidine kinase to the monophosphate. Not cytotoxic. Can be used topically, orally and intravenously
Bromovinyldeoxyuridine	More active than acyclovir against HSV-1 and varicella-zoster. Use is still experimental.
Ribavirin (Virazole)	Active in-vitro against a wide range of RNA and DNA viruses but little evidence of efficacy in clinical trials. Causes haemolytic anameia

DNA viruses

POXVIRUSES

— The largest viruses; just visible by light microscopy
— Have a predeliction for epithelial cells in which they produce
  characteristic eosinophilic inclusions in the cytoplasm
— Characteristic feature of the diseases is skin lesions; papules,
  vesicles and pustules (proliferative lesions in birds)
— Species-specific poxviruses exist for many animals and birds
— The poxviruses of man include:
  Variola
  — Smallpox ⎤
  — Alastrim ⎦ (now extinct)
  Vaccinia
  — Attenuated strain of cowpox used for vaccination against
    smallpox
  Paravaccinia
  — Milkers' nodes
  Cowpox
  Orf
  — Contagious pustular dermatitis of sheep
  — Occasionally causes skin lesions in man
  Molluscum contagiosum
  — Transmitted venereally

HERPESVIRUSES

— A large group of viruses infecting many animal species
— With the exception of Epstein–Barr virus, produce giant cells and
  intranuclear inclusions in tissue cultures
— Replication of viral DNA takes place within the nucleus where it is
  joined by viral capsid protein which migrates from the
  cytoplasm. Viral glycoproteins are incorporated into the nuclear
  membrane which forms the viral envelope as virions bud out
  through the nucleus
— After primary infection, herpesviruses sequester into sites from
  which they can cause recurrence at any time in the life of the
  host. Immunosuppression potentiates reactivation

Herpesviruses of man

1. Herpes simplex
   Epithelial lesions at site of entry. Local recurrences
2. Varicella-zoster
   ? Enters via upper respiratory tract
   Viraemia is followed by generalised vesicular eruption (chickenpox).
   Recurs as herpes zoster (shingles)
3. Cytomegalovirus
   Occasional cause of mononucleosis syndrome.
   Important cause of infection in immuno-suppressed patients.
   Commonest cause of transplacental infection.
4. Epstein-Barr virus
   Infectious mononucleosis (glandular fever).
   Occasional cause of lymphoproliferative disease after organ transplantation.

Diagnostic methods for herpesviruses

1. Cytology
   Scrapings of the base of the vesicular skin lesions of herpes simplex, varicella and herpes zoster stained with Giemsa yield squamous epithelial cells with ballooning degeneration, giant cells and intranuclear inclusions (Tzanck preparation).
2. Electron microscopy
   The crystalline structure of capsids of herpesviruses are readily recognised in exudate from skin lesions.
3. Culture — is of variable diagnostic utility:
   Herpes simplex: grows in any cell type to produce easily recognised CPE in 48–72 hours
   Varicella: CPE in human diploids, in 10–14 days
   Cytomegalovirus: CPE in human diploids, in 2–3 weeks
   Epstein–Barr: grows only in B lymphocytes, making them 'limmortal'. No CPE
4. Serology
   a. Complement fixation tests
      — produce relatively short-lived high titres in primary or recrudescent disease
      — are partially cross-reactive within the group
   b. Immunofluorescent tests
      — useful for detection of antigen (e.g. Herpes simplex in brain tissue) as well as antibodies
      — used to detect antibodies of various EBV antigens, using infected B lymphocytes
   c. Neutralisation tests
      — produce lifelong high titres. Useful for epidemiological studies

Herpes simplex

A universal virus of humans acquired by close personal contact. Causes vesicular lesions at the site of mucosal inoculation, as shown in Table 52.

Table 52

Site	Mode of transmission	Clinical features
Oral cavity	Kissing	Gingivitis, stomatitis. Recurs as labial cold sores.
Eye	Contamination with oral secretions	Keratoconjunctivitis. Recurs as dendritic ulcers.
Penis Vulva	Sexual intercourse	Vesicular lesions
Fingers	Handling of oral secretions	Paronychia (especially in nurses)
Skin	Contamination of abraded skin with saliva	Vesicular lesions, especially over buttocks and thighs, in wrestlers ('gladiatorial' herpes)

Primary infections in adults may be severe with rash lasting for 3 or 4 weeks.

In non-immune subjects, the virus travels along the sensory nerves and persists lifelong in the sensory ganglion, being later reactivated by fever, immune suppression, etc, to travel down the nerve fibres causing recurrent lesions.

HSV includes two serotypes which are acquired independently of each other and the rate is determined by the socio-economic status of the community, as follows:

HSV-1 In poorer communities, infection is acquired early in life. Almost 100% have antibodies by the age of 20. In wealthy communities, about 50% have antibodies by adulthood

HSV-2 Rarely acquired before puberty or in celibate groups such as nuns. About 20% of sexually-active adults have antibodies

A comparison of clinical and laboratory features of the two serotypes is given in Table 53.

Table 53

	HSV-1	HSV-2
Primary site	Lips 80–90%	Genitalia 80–90%
Encephalitis	Yes	No
Recurrences	Less frequent	More frequent
Neurotropism in mice	Less	More
Pock size on chorio-allantoic membrane	Small	Large
Plaques in chick embryo monolayers	−	+
Growth at 40°C	−	+
Heparin sensitivity	+	−
Syncytial giant cells in tissue cultures	−	+
Sensitivity to idoxuridine	+++	+

Complication of Herpes simplex infections
Encephalitis: follows primary infection or reactivation of HSV-I
Meningitis: occasionally complicates primary infection with HSV-2
Congenital infection: results from maternal viraemia with either HSV 1 or 2
Neonatal infection: is acquired from the birth canal during vaginal delivery
Erythema multiforme: may follow attacks of herpes simplex
Malignancy: There is an uncertain association between carcinoma of the cervix and HSV-2

Varicella-zoster

Chickenpox was first distinguished from smallpox by Heberden in 1802.

Primary infection is characterised by generalised vesicular rash.

Reactivation of the virus in dorsal root ganglia often years later causes herpes zoster (shingles) with vesicular rash in the distribution of a sensory dermatome.

Clinical features

1. Varicella
— is a highly infectious disease of childhood probably transmitted by the respiratory route.
— follows an incubation period of 12–17 days with fever followed by characteristic crops of vesicles more frequent on the trunk than the extremities (i.e. 'centripetal', as opposed to the 'centrifugal' distribution in smallpox)
— causes little systemic upset in children but is more severe in adults
— may be complicaticated by the following:
  Skin infection: may be caused by staphylococci or streptococci
  Encephalopathy: includes viral encephalitis (usually uncomplicated) and Reye's syndrome
  Aseptic meningitis
  Pneumonia: develops 1–6 days after the onset of rash producing cough, dyspnoea, hypoxia and a diffuse nodular infiltrate which may later calcify
  Congenital (transplacental) infection
  Bleeding disorders: a. thrombocytopenia
  b. disseminated intravascular coagulation
  Arthritis
  Overwhelming infection: may occur in immunodeficient patients, especially non-immune leukaemic children or following transplant

2. Herpes zoster
— is recrudescent varicella occuring in the dermatome of a latently infected sensory nerve ganglion
— occurs more frequently in old age and in the immunosuppressed
— causes pain followed by a vesicular rash which is morphologically similar to varicella
— is distributed amongst sensory root ganglia as follows:
  Thoracic: causes shingles on the trunk (*zoster* Greek = a belt), accounts for about half of the cases
  Cervical Lumbar each account for about 20% of cases
  Trigeminal nerve: about 15% of cases. Involvement of the 2nd division may cause keratitis, iridocyclitis, corneal ulceration
  Geniculate ganglion: Vesicles in external ear and anterior portion of tongue, facial nerve palsy (Ramsay–Hunt syndrome)
Complications:
Encephalomyelitis
— CSF pleocytosis is relatively frequent, prognosis is good.
Post-herpetic neuralgia
  Relatively frequent in the elderly.
Disseminated infection
— Results in vesicles appearing on skin outside the involved dermatome(s), usually in immunosuppressed.

Cytomegalovirus
— In normal hosts, infection with cytomegalovirus usually causes
  no symptoms
— In Western countries, about 60% of young adults have antibodies
— About 1% of women of childbearing years acquire antibodies
  annually
— Subclinical infection is followed by latent infection in various cell
  types, including circulating leucocytes. The virus is excreted in
  saliva, urine, semen, breast milk and cervical secretions
— Rate of excretion and circulating antibody titres are increased in
  the presence of immune deficiency induced by
  immunosuppressives, pregnancy and debilitation
— Modes of transmission:
  1. Close contact (probably accounts for most infection)
  2. Blood transfusion
  3. Organ transplant
  4. Transplacental

Clinical spectrum of infections
1. Mononucleosis syndrome
   — occurs in otherwise healthy young adults,
   — causing an illness similar to glandular fever but with less
     pharyngitis and lymph node enlargement
   — is associated with circulating atypical lymphocytes but unlike
     EBV infection, heterophil antibodies do not appear
   — may be associated with a mild hepatitis
2. Following cardiac bypass surgery or massive transfusion
   — causes mononucleosis syndrome
3. Congenital infection
   — follows maternal viraemia, usually associated with
     asymptomatic primary infection
   — is possibly the commonest cause of congenital infection with
     published rates of 0.5–2.5%
   — causes sequelae ranging from nil to microcephaly, blindness,
     mental retardation, spasticity and epilepsy
4. Guillain-Barré syndrome
5. Infection following organ transplant:
   — is acquired as follows:
     a. Primary infection acquired with the organ or via blood
        transfusion. If immunosuppression is maintained, these
        infections have a high mortality
     b. Recrudescent disease following reactivation of latent
        infection. Clinical manifestations range from nil to severe,
        depending upon the degree of immunosuppression

— is manifest clinically as follows:
   a. High swinging fever
   b. Leucopenia
   c. Interstitial pneumonitis
   d. Hepatitis
   e. Colitis
   f. Retinitis

Epstein-Barr virus
— Multiplies in B lymphocytes causing polyclonal growth stimulation and lymphoid hyperplasia
— In-vitro infection of lymphocytes causes them to multiply indefinitely ('immortalisation') and to express the latent genome only in the form of nuclear neoantigens (Epstein–Barr nuclear antigen)
— 12–25% of healthy adults shed EBV in pharyngeal secretions, probably following multiplication and release from epithelial cells
— Activation of T cells in response to B cell infection creates the characteristic atypical lymphocytosis seen in peripheral blood during infection

Clinical features
— incubation period of 30–50 days following pharyngeal exposure (kissing)
— more often subclinical in the first decade
— fever
— pharyngitis, palatal enanthem
— lymphadenopathy
— splenomegaly, hepatomegaly
— rash, especially following administration of aminopenicillins
— convalescence may be prolonged

Complications of EBV infection
Hepatitis: Jaundice occurs in about 5%.
Pneumonitis: in 2%
Neurological: signs occur in about 1%, as follows:
a. Aseptic meningitis
b. Encephalitis, myelitis, optic neuritis
c. Acute cerebellar syndrome
d. Guillain-Barré syndrome
e. peripheral neuropathy
f. Bell's palsy
Haematological
a. Haemolytic anaemia
b. Thrombocytopenia
c. Splenic rupture
Myocarditis, pericarditis

Manifestations in certain hosts
Burkitt's lymphoma
— is common in the malarious part of Africa. Strong association
with EBV markers
Nasopharyngeal carcinoma
— common in southern Chinese, is strongly associated with EBV
markers.
In male children with X-linked immune deficiency
a. Overwhelming infection
b. Agammaglobulinaemia
c. Lymphoma
In immunosuppressed
— especially follows the use of cyclosporin, causes polyclonal
lymphoproliferative disease, mostly extranodal

Diagnosis
1. Atypical lymphocytosis
   (also a feature of primary cytomegalovirus infection and present
   in smaller numbers in many viral infections)
2. Heterophil antibodies (Paul-Bunnell test)
   — are antibodies to sheep, beef or horse red cells NOT absorbed
   by guinea pig kidney
   — present in 90% of cases but sometimes not until 3rd or 4th
   week of infection. Less common in first decade
   — usually detected using commercial kits of formalised horse
   red cells
3. Antibodies to the various antigens of EBV, as shown in Table 54.

Table 54

Antigen	Symbol	Antibody response
Viral capsid	VCA	IgG antibodies are present at onset, persist lifelong IgM antibodies are present at onset, persist for 4–8 weeks. The best single test for diagnosis of recent disease
Nuclear antigen	EBNA	Antibody appears about 3 weeks after onset, persists lifelong
Early antigen (diffuse)	EA, anti-D	Appears in 70% of cases for 3–6 months
Early antigen (restricted)	EA, anti-R	Seen in protracted illness and in African Burkitt's lymphoma

ADENOVIRUSES
— Species-specific infections of many animals
— Produce cytopathic effects in tissue cultures, with swelling and clustering of infected cells which contain intranuclear inclusions
— Some human strains cause malignant tumours in baby hamsters
— Cause mucosal infections as well as colonising lymphoid tissue
— 37 serotypes are officially recognised
— Recently identified non-cultivatable types have been implicated in enteritis (serotypes 39 and 40)

Clinical syndromes
1. Pharyngoconjunctival fever
2. Coryza with fever
3. Pneumonia
4. Acute follicular conjunctivitis
5. Epidemic keratoconjuntivitis
6. Haemorrhagic cystitis
7. Possible association with intussuseption, mesenteric adenitis, diarrhoea

HEPADNAVIRUSES
A class recently created to include hepatitis B and related animal viruses which have the following characteristics:
1. Contain very small circular, partly single-stranded DNA molecules
2. Contain a DNA polymerase which repairs the single-stranded component of the DNA molecule following entry into the cell, making it fully double-stranded
3. Virions have a diameter of about 42 nm and an inner core of 27 nm
4. A lipid-containing envelope carries the surface antigen
5. Numerous particulate forms bearing the surface antigen may be present in the serum of infected subjects. May be spherical or filamentous, consist of lipid, protein and carbohydrate
6. Cause hepatitis which is commonly persistent
7. Similar viruses have been found in woodchucks, ground squirrels and Pekin ducks

Table 55 Antigens of hepatitis B virus

Surface antigen	HBs Ag	Major component of circulating particles (incomplete virions) Present in cytoplasm of infected hepatocytes Includes a group specific determinant as well as two pairs of subtypes (d,y and w,r) Antibodies to HBs Ag are protective
Core antigen	HBc Ag	Present mostly in nucleus of infected hepatocytes. Circulates only in complete virion (Dane particle). Includes DNA polymerase
e antigen	HBe Ag	Soluble 300,000 dalton protein found in the circulation during acute infection Constitutes a complex of antigens
Delta antigen	δ Ag	Defective RNA virus which superinfects HBV infections

(The use of these antigens in the diagnosis of HBV infections is described in the chapter on hepatitis)

PAPOVAVIRUSES

— The word is a contraction of the names papilloma, polyoma and vacuolating agent
— Small naked viruses with cubic symmetry, 45–55 nm particles
— Most are potentially oncogenic as judged by their ability to transform cells of various species in tissue culture
— Family includes two genera (Table 56)

Table 56

	Size	Haemagglutinin	Growth in tissue culture	Human species
Papillomavirus	52–54 nm	Absent	No	Human wart virus
Polyomavirus	45 nm	Present	Yes	JC and BK viruses, SV 40

Progressive multifocal leucoencephalopathy (PML)

— A rare disease occuring only after prolonged immunodeficiency (e.g. AIDS, lymphoma, myeloproliferative diseases) due to JC virus
— Causes widespread demyelination with little inflammatory response. Intranuclear inclusions inside oligodendrocytes contain virus particles
— Giant astrocytes which appear in late lesions have pleomorphic almost malignant-looking nuclei
— Progresses to death over 2–4 months

PARVOVIRUSES

Small, naked 20–30 nm particles. Cause disease in several species.

The association of parvoviruses with clinical disease has only recently been described in man:

1. Serum parvovirus
   a. Erythema infectiosum
      — also known as 'fifth disease', being the fifth of a series of exanthematous diseases of childhood delineated in the 19th century
   b. Aplastic crises in haemolytic disease
      — especially in sickle cell disease, following the multiplication of the virus in rapidly dividing bone marrow cells
2. Intestinal parvoviruses
   Gastroenteritis
      — Several agents, named geographically

RNA viruses

ORTHOMYXOVIRUSES

Medium-sized enveloped viruses the morphology of which has been intensively studied to reveal the following:

1. Envelope
— is composed of a double layer of lipid the inner surface of which is covered with a structural (membrane) protein
— is covered with two types of projecting surface glycoproteins:
Haemagglutinin
— Rod-shaped spikes which are the site for virus attachment to host cells. Contains type and strain specific antigens and is the protein most involved in antigenic variation
— Attachment to red cells is inhibited by antibodies
Neuraminidase
— Mushroom-shaped projections containing neuraminidase activity

2. Nucleocapsid
— consists of 8 discrete segments of ribonucleoprotein which are randomly incorporated into the virion during its maturation. Simultaneous infection of a cell with two different influenza viruses creates antigenic variation in the progeny (antigenic shift)
— includes 3 types of RNA polymerases
— possesses type-specific antigenicity on which the classification of 3 influenza viruses is based, viz.

Influenza A: causes pandemics associated with antigenic change
Influenza B: periodic epidemics, strong association with Reye's syndrome
Influenza C: is much less virulent than the above

Clinical features
1. Incubation period of 24—72 hours followed by abrupt onset of fever and myalgia
2. Sore throat, dry irritating cough
3. Complications include:

Pneumonia
a. Primary viral, is a rare sequel to influenza. Occurs at any age, has a high mortality
b. Bacterial, follows damage of bronchial mucosa. Usually due to *Streptococcus pneumoniae* and *Haemophilus influenzae* but *Staphylococcus aureus* pneumonia also has a common association with influenza

Reye's syndrome
Mostly follows influenza B.

Myositis

Myocarditis

Diagnosis
1. Virus culture
2. Detection of antigen in infected respiratory epithelial cells by immunofluorescence
3. Serology, using complement fixation test

Antigenic changes in influenza A
1. Antigenic *shift* probably results from recombination of RNA segments when two antigenic types of influenza infect the same cell, especially in birds. Results in major antigenic change of both surface components
2. Antigenic *drift* results from mutations which cause minor antigenic change in haemagglutinin.
   Antigenic changes are designated according to serotype of the haemagglutinin (H) and the neuraminidase (N) as illustrated in pandemics which have been recorded as shown in Table 57.

PARAMYXOVIRUSES
— Enveloped helical RNA viruses three of which have haemagglutinins.
— Unlike myxoviruses, genetic recombination does not occur.
— Four viruses are pathogenic for man (Table 58).

Table 57 Antigenic changes in influenza A

Year	Symbol	Popular designation	Characteristics of epidemics
1889	?H2N2		
1900	?H3N2		
1918	HswN1	'Swine'	21 000 000 deaths recorded worldwide. The largest epidemic in history
1929	HON1		Not pandemic. Was transmitted to ferrets in 1933 and grown in chick embryos in 1940
1946	H1N1		
1957	H2N2	'Asian'	Severe. Staphylococcal pneumonia was a frequent sequel
1968	H3N2	'Hong Kong'	Moderately severe. Elderly (i.e. those born before 1918) were immune
1976	HswN1	Fort Dix	Produced a small outbreak in New Jersy but no epidemic.
1977	H1N1		Mild. Population born before 1957 are immune.

Table 58 Classification of paramyxoviruses

	Haemagglutinin	Neuraminidase	Site of replication	Sero types	Syndromes
Parainfluenza	Present	Present	Cytoplasm	1,2,3,4A and 4B	Croup Lower respiratory tract infection
Mumps	Present	Present	Cytoplasm	One	Parotitis Orchitis
Measles	Present	Absent	Cytoplasm and nucleus	One	Measles
Respiratory syncytial (RSV)	Absent	Absent	Cytoplasm	One	Bronchiolitis Pneumonia

Diagnostic methods
1. Culture (Table 59)

Table 59 Culture of paramyxoviruses

Virus	Cell type	Cytopathic effect
Parainfluenza	Monkey kidney (not others)	Type 2 produces CPE Haemagglutinate guinea pig RBC
Mumps	Monkey kidney HeLa cells Hen's eggs	Rounding of cells Multinucleate syncytia Eosinophilic inclusions Haemagglutinate chick RBC
Measles	Monkey kidney	Multinucleate giant cells
RSV	Cell lines	Characteristic refractile syncytium formation

2. Detection of antigen in infected cells (immunoflourescence)

3. Detection of antibodies
Two antigen preparations are available:

'S' (Soluble) antigen — Nucleocapsid
        — Antibodies appear early and wane quickly.

'V' (Viral) antigen   — Surface glycoproteins with haemagglutinating and neuraminidase activity
        — Antibodies appear later and persist

Although there is no antigen common to the whole group, heterologous antibody responses to other paromyxoviruses follow infection with one member of the group.

Methods for detecting antibodies are as follows:

Complement fixation
Most frequently used. Detects high titres for short periods after infection

Haemagglutination inhibition
Uses 'V' antigen, high titres are prolonged

Neutralisation
Sensitive but cumbersome to perform

Measles
— A ubiquitous infection of early childhood spread by aerosol formation and extremely contagious. Without vaccination most children have acquired the infection by the age of five
— Incubation period is 10–15 days during which time viraemia leads to virus multiplication in respiratory epithelium and skin
— Begins with fever, conjunctivitis, rhinorrhoea and cough
— A buccal enanthem appears around the parotid duct orifices (Koplik's spots)
— 2–4 days later, the typical maculopapular rash appears first on the face then spreading over the whole body

Complications
1. Mucosal infections
   — including conjunctivitis, otitis media, bronchopneumonia
2. Diarrhoea
   — occurs mainly in the tropics, may be due to mucosal involvement by measles virus or bacterial infection
3. Post-infectious demyelinating encephalitis
4. Subacute sclerosing panencephalitis (SSPE)
   — is due to persistent latent infection of neurones by apparently defective measles virus
   — develops months to years after apparent recovery
   — results in intellectual deterioration, myoclonic movements, characteristic EEG changes and death in months
   — is recognised histologically by the presence of cytoplasmic inclusions
5. Giant cell pneumonia
   — is a rare form of progressive measles infection of the lung in children with leukaemia
6. Suggested (but unproven) association with multiple sclerosis

Mumps
The origin of this Anglo Saxon word is obscure but probably refers to the facial expression associated with parotid enlargement.

After an incubation period of 12–21 days, salivary glands enlarge and become painful, especially during eating. Fever and malaise may be associated.

Complications (more frequent in adults) include:

Aseptic meningitis	Relatively common and mild
Orchitis	Occurs in up to 25% of postpubertal cases, usually unilateral, causes severe pain and swelling
Pancreatitis	Occurs in up to 10% of adults
Oophoritis Arthritis Thyroiditis	are very rare sequelae
Encephalitis	Demyelinating, rare

Respiratory syncytial virus (RSV)
— Although classed as a paramyxovirus, RSV possesses neither a haemagglutinin nor a neuraminidase
— The name of the virus reflects the highly characteristic cytopathic effect seen in tissue culture
— RSV is the most frequent cause of bronchiolitis in early childhood with peak incidence occurring at 3 months of age
— In winter, RSV accounts for most paediatric hospital admissions due to lower respiratory tract infection
— There is some evidence to suggest that the severe pneumonic effects of RSV are mediated by reactions between the virus and maternally — acquired antibodies. Killed vaccines do not protect
— Re-infections in older children and adults cause only upper respiratory tract symptoms

Parainfluenza virus
— A paramyxovirus which infects animals as well as man
— Most children have acquired infection by the age of five
— Includes four serotypes with varying clinical and epidemiological features, as follows:

Type 1 } Cause croup in infants, often in winter epidemics
Type 2 }

Type 3   Causes bronchiolitis and pneumonia under the age of 6 months as well as croup in older infants. Virus circulates throughout the year

Type 4   Causes only minor respiratory symptoms

PICORNAVIRUSES
A large family of animal viruses characterised by small size and naked icosahedral symmetry (*pico* = very small; RNA, nucleic acid type).
  Two genera commonly infect man:
1. Enteroviruses
  — Stable at pH 3–10, resist gastric acid and inhabit the gastrointestinal tract
2. Rhinoviruses
  — Unstable below pH 6, optimal temperature of growth is 33°C. Inhabit upper respiratory tract. One of several virus groups responsible for common colds.

Enteroviruses
— were previously classified according to antigenic relationships and host range.
— consist of distinct immunotypes the newer identified of which have simply been numbered rather than named.

Table 60 Types of enteroviruses

Species (immunotypes)	Number in the group	Host range
Polioviruses	3	Paralysis in primates
Coxsackieviruses A	24	Poor growth in cell cultures Generalised myositis in suckling mice
Coxsackieviruses B	6	Encephalitis in suckling mice
Echoviruses	34	Nonpathogenic for mice and primates
Enteroviruses 68–71 (unnamed)	4	
Enterovirus 72 (Hepatitis A virus)	1	Hepatitis in man and primates

Clinical syndromes
  1. Viraemic illness (no   All enteroviruses
     organ involvement)
  2. Meningitis   Most enteroviruses
  3. Encephalitis   Most enteroviruses (but rare)
  4. Poliomyelitis   Poliovirus, occasionally others
  5. Pharyngitis   Coxsackie A
  6. Rash   Coxsackie A
  7. Pericarditis Myocarditis   Coxsackie B
  8. Pleurodynia, myositis   Coxsackie B
     (Bornholm disease)
  9. Orchitis   Coxsackie B
 10. Conjunctivitis   Enterovirus 70
 11. Hepatitis   Hepatitis A virus, rarely
                                 coxsackieviruses

Diagnostic methods
— pose a problem in this group
1. *Electron microscopy* — is unsatisfactory because the infections
   are not superficial and the viruses are small and featureless
2. *Culture* — of CSF, throat, faeces but note that:
   a. incidental faecal carriage of enteroviruses is common in
      young children.
   b. Coxsackieviruses A do not usually grow in tissue culture
3. *Serology* has limited value because each of the immunotypes is
antigenically distinct and it is not feasible to test each type
separately

Epidemiological characteristics
1. Immunity to each enterovirus is type-specific
2. Highest infection rates occur under one year of age although symptomatic disease is more common in older children. Symptomatic infections in the elderly are rare
3. More prevalent in summer and autumn. The frequency of faecal carriage increases in warm climates
4. Spread is mostly from person to person with marked clustering within families
5. Man is primarily infected by ingesting faecally contaminated food. Infection is more prevalent in lower socio-economic groups
6. A good proportion of infections even during an epidemic are asymptomatic. Faecal carriage after infection is occasionally prolonged

RHABDOVIRUSES

— *Rhabdos* (Greek) = a rod
— A group of viruses with a bullet-like shape in electron microscopy

Rabies
— Enzootic in many countries, in terrestrial warm-blooded animals
— Dog bites are responsible for most cases. In Western Europe, red foxes are the main transmitters of the disease
— Rabid animals begin secreting virus in their saliva several days before the onset of symptoms

Modes of transmission
1. Animal bites, usually by dogs
2. Contamination of scratch wounds by dog saliva
3. Laboratory accidents
4. Corneal transplants from unrecognised victims of the disease
5. ? Inhalation, in bat-infested caves

Pathogenesis:
1. Virus probably multiplies in muscle cells near the site of the bite
2. After a long latent period, spread to the central nervous system occurs along the axons of peripheral nerves
3. Antibody responses occur late

Clinical features
Include:
— incubation period of 18–60 days, longer if the bite is on an extremity rather than the head
— prodrome of malaise, headache, fever, paraesthesia at the site of exposure
— acute neurological signs, including agitation, hallucinations, seizures, pharyngeal spasm provoked by drinking
— coma lasting hours to months. Only occasional recovery has been documented

Diagnostic methods
1. Isolation of virus from saliva, CSF, brain or urine
2. Detection of antibodies in serum during the second and third week of the illness, or later in CSF
3. At autopsy, brain examination reveals perivascular inflammation of grey matter and in most cases characteristic cytoplasmic inclusions (Negri bodies) especially in neurones of the hippocampus and Ammon's horn

Management
1. Washing of the wound with soap and water
2. Human rabies immune globulin
3. Antirabies vaccine (human diploid)

REOVIRUSES

(REO = Respiratory enteric orphan viruses; so named because the originally isolated orthoreoviruses appeared to be unassociated with disease.)
— The only double-stranded RNA viruses to infect man. 60–80 nm naked particles with cubic symmetry
— Three genera
1. Orbiviruses
  — Arthropod-borne infections, three are pathogenic for man
2. Rotaviruses
  — Common cause of diarrhoea in infants and animals
3. Orthoreoviruses
  — Cause coryza

Rotaviruses
— 70 nm particles with a double-shelled capsid and characteristic electron microscopic wheel-like appearance
— The genome consists of 11 distinct segments of double-stranded RNA capable of reassortment (like influenza)
— Rotaviruses have been detected in the stools of numerous infant animals, all having an antigenically distinct outer capsid
— Human strains grow poorly or not at all in tissue culture
— Cause winter gastroenteritis mostly in age group 6 months to 2 years
— Damage the absorptive epithelial cells of upper small intestine causing enzyme deficiency
— Infection is characterised by vomiting, low fever, diarrhoea
— In developed countries rotavirus is the commonest cause of infantile dehydration requiring hospital admission
— More than 90% of infants have antibodies by the age of 2
— May spread through neonatal nurseries but usually without clinical manifestations

Diagnosis
1. Electrom microscopy
2. Antigen detection (ELISA, RIA, IFA)

Treatment
1. Rehydration
2. ? breast milk, hyperimmune cows' milk
3. Oral pooled gammaglobulin

CORONAVIRUSES

— 'Corona' describes the crown-like surface projections
— Cause numerous mucosal infections in domestic animals
— The coronaviruses of man are difficult to culture and have thus far only been associated with common colds

TOGAVIRUSES

— Small (40–90 nm) enveloped RNA viruses with cubic symmetry.
— Classified according to size and mode of transmission as follows:
Alphaviruses: Previously classified as group A arboviruses
Flaviviruses: Previously classified as group B arboviruses
Smaller than alphaviruses. Includes yellow fever and dengue.
Rubivirus: A genus created for rubella, a non-arthropod-borne, togavirus

Rubella

— Closely related to alphaviruses but not arthropod-borne.
— Most common in spring. Epidemics occur every 6–9 years as susceptible population increases. Major epidemics have occurred approximately every 30 years (before the introduction of vaccination)
— Spread by respiratory secretions but only moderately contagious
— About 80% of adults are seropositive
— Most infection is probably subclinical especially in children
— Incubation period 12–23 days (average, 18)
— Symptoms include cervical lymphadenopathy, low fever and rash which appears as immunity developes
— Asymptomatic reinfections can occur almost always without viraemia or risk of transplacental spread

Complications
Arthritis
— especially in adult women
— occurs mostly in fingers, wrists and knees, may persist for weeks
Bleeding
— occasionally accompanies thrombocytopenia
Congenital rubella
— is a consequence of rubella acquired in early pregnancy
— causes frequent fetal death and congenital defect

Diagnosis
1. Virus isolation — is time consuming, requires unusual cell types
   (African green monkey kidney, rabbit)
2. Serology — establishes the diagnosis by detection of antibodies
   in the IgM class for up to 30 days after development of rash.
   Techniques include:
   a. Haemagglutination inhibition (HAI)
      — is most frequently used but inhibitors of red cell
        agglutination must be removed from serum
   b. Enzyme-linked immunosorbent assay (ELISA)
      — is now replacing HAI
   c. Single radial haemolysis (SRH)

ARENAVIRUSES

— Enveloped viruses with club-shaped glycoprotein projections
  and containing variable numbers of electron dense granules in
  the interior of the virion (arenosus, Latin, sandy), representing
  host ribosomes
— Viral nucleic acid consists of four large pieces of DNA and several
  small ones
— Rodents are the natural hosts of arenaviruses, maintained by
  chronic infection. Humans are infected by accidental contact with
  urine
— Four viruses cause infection in man (Table 61).

BUNYAVIRUSES

— 90–100 nm particles, enveloped with surface projections
— Contain three circular ribonucleoprotein strands capable of
  genetic reassortment. Virion includes a RNA polymerase
— The several groups of viruses include:
  California encephalitis: Commonest causes of encephalitis in
                            Midwest USA
  Phlebotomous (sandfly) fever: Cause mild, self-limited fever in
                                 Mediterranean and Central
                                 Asia
  Rift Valley fever: Disease of sheep and cattle in East Africa now
                      occurring in humans. Causes encephalitis
                      and macular exudates.
  Hantaviruses: Haemorrhagic fever with renal syndrome (HFRS)

Table 61 Arenaviruses of man

Virus	Distribution	Clinical features
Lymphocytic choriomeningitis	? world-wide	Flu-like illness may be followed by aseptic meningitis, encephalomyelitis
Lassa fever	a broad belt of West Africa	Clinical features range from mild to severe with multiple organ involvement Can be transmitted from person to person
Argentine haemorrhagic fever	farmland west of Buenos Aires	Typical haemorrhagic fever
Bolivian haemorrhagic fever	Beni province of Bolivia	Typical haemorrhagic fever

Hantaviruses
— Includes the agents of Korean haemorrhagic fever, Crimean haemorrhagic fever, Omsk fever and nephropathia epidemica and other viruses causing nephropathy across much of the Eurasian land mass
— Cause haemorrhagic fever with renal syndrome (HFRS), recently termed 'muroid virus nephropathies' (Gajdusek 1982)
— Asymptomatic infection of wild rodents in Asia, eastern Europe Scandanavia and North America. Transmitted to man by excretions
— Human infections occur mostly in rural areas but laboratory personnel who handle rats and mice are at special risk

Clinical
— Begins with abrupt fever, facial flush, widespread petechiae
— Proteinuria occurs after 3–5 days, accompanied by thrombocytopenia and leucocytosis
— Hypotension may follow, together with oliguria, haematuria and uraemia

ARBOVIRUSES

— ARBO = Arthropod-borne
— Viruses transmitted between susceptible vertebrates and arthropods, multiplying in the tissues of both hosts and passed on by the bite of the arthropod
— Includes over 400 viruses throughout the world
— Most common in tropical and rural areas where mosquito and animal reservoirs are plentiful
— Viruses multiply in the reticuloendothelial system and vascular endothelium

Clinical features
— Include:
1. Fever, malaise, myalgia, headache
2. Transitory skin rash
3. Localisation in certain tissues to cause encephalitis, arthritis, renal failure, hepatitis
4. Haemorrhage

Diagnosis
1. Isolation using mice or tissue culture is usually impractical
2. Serology
   IgM antibodies in haemagglutination-inhibition test
   This grouping is *ecological*, not taxonomic, and several classes of enveloped RNA viruses are involved:

1. Togaviruses   ⎫   Eastern equine encephalitis
   Alphaviruses   ⎬   Western equine encephalitis
   (40–50 nm)   ⎭   Venezuelan equine encephalitis
   Flaviviruses   ⎫   Yellow fever
   (50–70 nm)   ⎪   Dengue
   extensive   ⎪   St Louis encephalitis
   serological   ⎬   Japanese encephalitis
   cross   ⎪   Murray Valley encephalitis
   reactions   ⎭   Louping ill
2. Orbiviruses   Colorado tick fever virus
3. Bunyaviruses   California encephalitis group
4. Rhabdoviruses   Five rabies-related viruses may rarely cause disease in man

RETROVIRUSES
— Enveloped particles with cubic symmetry
— Possess a reverse transcriptase which synthesises DNA from the viral RNA template
— Tissue culture infection causes the cells to transform
— Bud from the surface of infected cells
— Cause malignancy by one of two possible mechanisms
   1. Oncogenes can be incorporated into the retrovirus genome
   2. Retrovirus DNA which does not contain oncogene activity enters the host DNA, activating neighbouring pro-oncogene
   Numerous retroviruses have been found to be the cause of malignancy in several animal species, viz,
Birds: Avian sarcoma, leucosis and myeloblastosis viruses
Mice: Mammary tumour, murine leukaemic and sarcoma viruses
Cats: Feline leukaemia and sarcoma viruses
Cattle: Bovine leukaemia viruses
   Human retroviruses have recently been discovered:

1. Human T-cell leukaemia virus (HTLV)
Isolated from cases of adult T-cell leukaemia in Japan, some American Whites, Black West Indians and parts of South America and Africa.

Causes leukaemia characterised by lymphadenopathy, skin involvement, hypercalcaemia and a leukaemoid blood picture with lymphocytes of bizarre morphology and a mature membrane phenotype.

2. Lymphadenopathy-associated virus (LAV), HTLV III
Retroviruses obtained from lymphocytes of male homosexuals with lymphadenopathy or acquired immune deficiency syndrome (AIDS). LAV specifically infects the 'helper' lymphocyte subset (characterised by the presence of the T4 surface marker)

AIDS
AIDS was first described in male homosexuals in New York and San Francisco in 1981 and has the following characteristics:
1. Occurs most frequently in:
    a. male homosexuals
    b. drug addicts
    c. recipients of blood and its products (especially haemophiliacs)
    d. certain countries (Haiti, West Africa)
    Mode of transmission appears to be similar to hepatitis B.
2. Incubation period is long (1–3 years)
3. Most patients have a prodrome of lymphadenopathy, weight loss, fevers and diarrhoea. Many will not develope AIDS
4. Patients with AIDS are profoundly immunosuppressed and prone to candidiasis, persistent herpes simplex, *Pneumocystis* pneumonia, cerebral toxoplasmosis, progressive multifocal leucoencephalopathy, cryptosporidiosis and Kaposi's sarcoma
5. Investigation reveals diminished 'helper' lymphocytes (the T4:T8 ratio is inverted) and antibodies to HTLV III

FILOVIRUSES
Enveloped RNA viruses with unique morphology, as follows:
a. The helical capsid has cross striations at 5 nm intervals and is surrounded by a close fitting envelope from which project 10 nm spikes
b. Elongated forms occur with branching or bizarre shapes
    The group is represented by two serologically distinct viruses, Marburg and Ebola viruses, both causes of severe haemorrhagic fever and both originating in Africa.
    The natural hosts and epidemiology are not known.
    Characteristics of these infections include:
1. Incubation period of 4–9 days
2. Rapid onset of fever, arthralgia, conjunctivitis, soon followed by vomiting and diarrhoea, leucopenia and thrombocytopenia

3. After a further day or so, rash and skin haemorrhages appear, associated with involvement of the brain, liver and kidneys
4. Persistent infection and viral excretion may occur

Diagnosis
1. Culture in continuous cell lines, identified by immunofluorescent antibodies or filamentous forms in electron microscopy
2. Rising antibody (IFA) titres

Fungi

— *Sphongos* (Greek) = a sponge
— *Eukaryotic*, spore-bearing, lack chloroplasts
— Filamentous and branched somatic structures (*hyphae*) which
  spread over food sources. A mass of hyphae is called mycelium
  (*mykes* (Greek) = a cap or a mushroom)
— Cell walls contain chitin and/or cellulose, resembling plants
— Reproduce by sexual (*meiotic*) or asexual means
  The *systematic classification* of fungi is based mainly on the
nature of sexual spores but has little relevance to fungi of medical
interest (Table 62).

Table 62

	Hyphae	Asexual spores	Sexual spores
1. Phycomycetes	Non-septate	Held in a bag (sporangium)	Oospores Zygospores
2. Ascomycetes	Septate	Bud off the ends of special hyphae (conidiophores)	Ascospores (8 together in a sac)
3. Basidiomycetes	Septate	May be formed by fragmentation of mycelium	Basidiospores (off the ends of basidia)
4. Fungi imperfecti	have no 'perfect' (i.e. sexual) state and therefore cannot be classified. Includes many fungi of medical interest		

Asexual methods of reproduction include:
1. Fragmentation of mycelium
2. Arthrospores
3. Chlamydospores
4. Budding
   Growth
— is best at 20–30°C (some will not grow at 37°C)
— optimal pH is 6.0, at high humidity
— Fungi can synthesise protein in the presence of carbohydrates
— may result in the production of numerous toxic compounds
  (mycotoxins) some of which may be ingested by man

119

Identification of fungi is based on:
1. Colonial appearance
2. Nature of asexual spores
3. Biochemical tests (mostly for yeasts)

The *medical classification* of fungi is best made on the morphology of vegetative forms (see Table 63).
1. Moulds
   — grow as hyphae and reproduce by various kinds of spores
   — form large filamentous colonies on artificial media
2. Yeasts
   — grow as single cells and reproduce by budding
   — form compact creamy colonies on artificial media
   — may form elongated cells (pseudomycelium) or true hyphae
3. Dimorphic fungi
   — grow as hyphae or yeasts, according to cultural conditions, viz
      a. at 22°C on Sabaroud's medium or in the   form hyphae
         soil
      b. at 37°C on blood agar or in tissues   form yeasts

A *clinical classification* of the diseases caused by fungi is as follows:
1. Superficial
   a. Tinea: Dermatophytes
             Tinea versicolor
   b. Candidiasis
2. Subcutaneous
   a. Chromoblastomycosis
   b. Sporotrichosis
   c. Mycetoma (Maduramycosis)
3. Systemic
   a. Aspergillosis
   b. Blastomycosis
   c. Candidiasis (occasionally)
   d. Coccidioidomycosis
   e. Cryptococcosis
   f. Histoplasmosis
   g. Mucormycosis
4. Mycotoxic
   a. Mushroom poisoning (jaundice)
   b. Ergotism (gangrene, psychosis)
   c. Drunken bread (ataxia, diarrhoea)
   d. 'Magic mushrooms' (hallucinations)

CANDIDIASIS

— is caused by *Candida* species, 4–6 µm yeasts which form part of the normal mucosal flora of man
— is most frequently due to *C. albicans* which is distinguished from other *Candida* species of man as follows:

C. albicans	Rapid (2 hour) formation of germ tubes in serum
	Chlamydospore formation
C.stellatoidea	Chlamydospore formation
C.guilliermondii	These species are distinguished by their
C.krusei	patterns of assimilation and fermentation
C.parapsilosis	of carbohydrates
C.tropicalis	
C.pseudotropicalis	
C.glabrata	Unlike other *Candida* species, *C. glabrata* does not produce pseudomycelium
	Also known as *Torulopsis glabrata*

Factors predisposing to candidiasis
1. Prior antibiotic therapy
2. Diabetes mellitus
3. Immune defect
   a. *immunosuppression*— steroids, cytotoxics
   b. *chronic mucocutaneous candidiasis (CMC)— a rare cause of persistent candidiasis associated with failure of T-cell lymphocytes to respond to stimulation with Candida* antigen and autoimmune endocrinopathies

Clinical manifestations of candidiasis

1. Superficial
Thrush
— White patches of pseudomembrane on the tongue and buccal mucosa which when scraped, leave a painful and raw bleeding surface
— Microscopy of scrapings reveals masses of yeasts, hyphae and pseudohyphae
Oesophagitis
— About half the cases are associated with thrush. Diagnosed by endoscopy or barium swallow
Vaginitis
— The most common cause of vaginitis. High circulating oestrogen levels (pregnancy, oral oestrogens) predispose
— Causes thick discharge and intense pruritis
Intertrigo
— Candidiasis at sites of skin approximation (crural folds, webs of hands and feet, beneath pendulous breasts, perianal region)

Table 63 Fungal infections of man

Fungus	Disease	Habitat, predisposing factors
MOULDS		
Dermatophytes	Ringworm (Tinea)	Most are distributed wordwide, some are primarily animal pathogens transmitted to man
Aspergillus species	Aspergillosis	Ubiquitous aerial contaminants, only invade compromised hosts
Mucor Rhizopus Absidia	Mucormycosis	Ubiquitous aerial contaminants, cause invasive sinopulmonary disease in uncontrolled diabetes and in neutropenia
Entomophthoraceae	Subcutaneous phycomycosis	Environmental organisms of Africa, India and S.E. Asia, causing subcutaneous granulomas
Various environmental fungi	Keratomycosis Maduramycosis	Maduramycosis occurs most frequently on the feet in tropical countries
Dematiacious fungi	Chromoblastomycosis	Dark pigmented fungi more frequently seen in tropical countries
YEASTS		
Candida species	Candidiasis	Normal human surface flora
Cryptococcus neoformans	Cryptococcosis	Soil saprophyte, multiplies in bird droppings. 50% of patients are immunocompromised
Malassezia furfur	Tinea versicolor	Lipophylic skin yeast which causes skin lesions in areas of high humidity

DIMORPHIC

Histoplasma capsulatum	Histoplasmosis	Soil organism which multiplies in bird and bat droppings. Very high incidence of human infection in the Mississippi valley. Portal of entry is the lung
Blastomyces dermatitidis	Blastomycosis	Environmental organism, most common in North America. Portal of entry is the lung
Coccidioides immitis	Occidioidomycosis	Soil organism of desert areas of California and Mexico. Portal of entry is the lung
Sporothrix schenckii	Sporotrichosis	Worldwide soil organism, usual portal of entry is the skin
Paracoccidioides brasiliensis	South American blastomycosis	Environmental organism restricted to South America. Portal of entry is the lung

Balanitis
— Thrush-like penile rash acquired through sexual intercourse with partner who has vaginal candidiasis
Paronychia
— Candida, along with bacteria, causes infection around the nails of subjects whose hands are frequently immersed in water ('barmaids' finger').
Napkin eruption
— Follows maceration of infant perineal skin by wet napkins.
Management of superficial candidiasis:
a. Exclude predisposing factors (especially diabetes)
b. Prevent the emergence of candidiasis in immunodeficient patients by use of topical polyenes (nystatin, amphotericin, natamycin)
c. Treat with topical polyenes or imidazoles

2. *Invasive candidiasis*
Predisposing factors:
a. Venous access
   — Cannulation
   — Intravenous drug abuse
b. Prolonged neutropenia
   — Portal of entry may be gastro-intestinal tract or venous line.
c. Other sites of cannulation
   — Renal tract
   — Peritoneal cavity
Sites of involvement:
Endocarditis
— Usually occurs on prosthetic valves
— High mortality without surgery
Retinitis
— Fluffy white chorioretinal lesions which may be detected days after an episode of candidaemia.
— May cause permanent blindness
Arthritis, osteomyelitis
— Haematogenous, especially in drug addicts
Peritonitis
— Follows peritoneal dialysis, abdominal surgery or ruptured viscus
Pulmonary
— Usually follows haematogenous seeding
— Microabscesses may not be seen on X-ray
Cerebral
— Microabscesses or meningitis, usually in leukaemics
Renal
— Ascending infection via catheter, or descending infection following candidaemia, occasionally obstructing the ureter ('fungus ball')

CRYPTOCOCCOSIS ('TORULOSIS')

— *Cryptococcus neoformans* is a yeast, 4–6 μm in diameter, surrounded by a large polysaccharide capsule
— Occurs worldwide in pigeon droppings and soil. Inhalation of contaminated dust results in pulmonary infection, usually circumscribed
— Progressive lung infection or disseminated infection occasionally follows, associated in about half the cases with immune deficiency or steroid administration
— Clinical manifestations include:

1. Lung
   a. Coin lesion ('toruloma') seen on chest X-ray. Must be differentiated from other lesions, including neoplasm, with a similar radiological appearance
   b. Pulmonary infiltrate which may simulate malignancy
2. Meninges
   Meningitis is the most frequent manifestation of cryptococcosis. Subacute onset
3. Other sites
   Skin, bone

Diagnosis of cryptococcosis

1. *CSF examination*—reveals pleocytosis, mostly lymphocytes, decreased glucose and capsulated yeasts, best seen in India ink preparation. Culture usually yields the organism but the polysaccharide capsular compound can be detected by latex agglutination test in virtually all cases
2. *Sputum examination*— of patients with pulmonary infiltrates may reveal capsulated yeasts in gram stain. Culture on selective media may yield *C. neoformans*
3. *Circulating capsular antigen* can be detected in many patients with meningitis and some with pulmonary infiltrates
4. *Lung biopsy* or *aspirate*—is required to establish the diagnosis of pulmonary toruloma

ASPERGILLOSIS

— *Aspergillus* species are ubiquitous environmental moulds the spores of which are present in air and are therefore continuously inhaled
— Infection only occurs in the presence of anatomical or immunological defect
— Clinical manifestations of lung disease include:
   1. Aspergilloma
   — 'Fungus ball' of the lung, follows colonisation of preformed cavities, especially of apices
   — May cause catastrophic haemoptysis
   — Antibodies to *Aspergillus* extracts are detected in the circulation

2. Hypersensitivity-induced disease
   Clinical and pathological entities include:
   a. allergic bronchopulmonary aspergillosis
   b. mucoid impaction of the bronchi
   c. bronchocentric granulomatosis
3. Invasive aspergillosis
   — Occurs in the presence of immunosuppression or
     neutropenia. Invades vessels producing segmental opacities,
     may disseminate to brain, heart and liver.
   — Almost impossible to diagnose without lung biopsy.

DERMATOPHYTES

Mycelial fungi, mostly imperfect, which penetrate and parasitize
keratin but which are inhibited by host factors from invading
subcutaneous tissues.

Table 64 Common types of ringworm

Tinea capitis (scalp)	Microsporum audouinii Microsporum canis	Primarily a disease of prepubertal children. M. canis is transmitted from cats and dogs Treated with oral griseofulvin
Tinea pedis (athlete's foot)	Trichophyton mentagrophytes	Clinical appearance includes: a. Scaling and fissuring between 4th and 5th toes
	Epidermophyton floccosum	b. Hyperkeratosis of plantar surface of foot, often chronic
	Trichophyton rubrum	c. Vesiculopustular lesions on plantar surface
Tinea corporis (trunk)	Various	Involves hairless skin, more common in tropics
Tinea cruris (groin)	Various	Predominantly in males
Tinea unguium (nails)	T.rubrum T.mentagrophytes	May involve distal or proximal nailbed, lifting and thickening the nail

CAUSES OF INTERTRIGO (inflammatory diseases of body skin folds)

1. Candidiasis
2. Ringworm (dermatophytes)
3. Erythrasma
4. Dermatitis
5. Flexural psoriasis

ANTIFUNGAL AGENTS

1. Topical

Gentian violet Whitfield's ointment	Time-honoured compounds which may still be useful in some situations
Undecenoic acid Tolnaftate	Used for dermatophyte infections, NOT candidiasis
Polyenes	Amphotericin, nystatin, natamycin Used for candidiasis, NOT dermatophytes
Imidazoles	Clotrimazole, miconazole, econazole Useful for tinea and candidiasis

2. Systemic (Table 65)

Table 65

Class	Agents	Indications
Polyenes	Amphotericin B	Invasive candidiasis Cryptococcosis Invasive dimorphic fungi
Base analogues	Flucytosine	Cryptococcosis Invasive candidiasis
Imidazoles	Miconazole (i.v.) Ketoconazole (p.o.)	Histoplasmosis Chronic mucocutaneous candidiasis Invasive candidiasis Dermatophytes
Griseofulvin		Accummulates in skin, inhibiting dermatophytes Used for tinea of the scalp and nails

Normal flora

The microbes normally present on the body surfaces are established in the weeks after birth. Consequences:
1. Interference with colonisation by pathogens
2. Stimulate formation of antibodies which may cross react with the surface components of potential pathogens
3. Vitamin production (Vitamin K2, in the gut)
3. Consumption of vitamins (in the gut)
4. Metabolic transformation of compounds, including the production of carcinogens (? causative role in carcinoma of gastrointestinal tract)

FLORA PRESENT AT VARIOUS SITES

1. Nasal vestibule
Staphylococcus epidermidis are the dominant species, as on skin
Diphtheroids
Staphylococcus aureus is present in 20–30% of normal adults.

2. Pharynx
Alpha haemolytic streptococci are the dominant *aerobic* species
Staphylococci
Branhamella catarrhalis
Diphtheroids
Anaerobes are the dominant species of the oral cavity
Haemophilus influenzae and other species are present in most children and adults
Streptococcus pyogenes is present in 10–20% of children
Streptococcus penumoniae is present in the upper respiratory tract of most children and adults
Candida species
Mycoplasma species
Viruses: adenoviruses, Epstein–Barr virus are excreted for prolonged periods after infection, which is usually subclinical

3. Gastrointestinal tract

a. Stomach and upper small bowel is sparsely colonised with pharyngeal organisms

Absence of gastric acidity promotes the growth of oral flora

Small bowel stasis promotes the growth of anaerobes and coliforms, resulting in malabsorption (blind loop syndrome)

b. Colon is heavily colonised with:

Anaerobes (approximately 10^{11} organisms per gram)

Enterobacteriaceae: *Esch. coli* predominates

Streptococci, lactobacilli

Smaller numbers of *mycoplasmas, yeasts*, and *viruses* may be present. 1–5% of adults may harbour *salmonellae, amoebae*.

Factors influencing colonic flora:

a. *Diet*

Undigested fibre is degraded and used as an energy source by colonic bacteria resulting in an increase in stool bulk

b. *Antibiotics*

Some antibiotics are imperfectly absorbed, some parenterally administered antibiotics are secreted into bile. Elimination of colonic anaerobes results in overgrowth of aerobic gram negative rods and yeasts

c. *Motility*

Increased peristalsis and diarrhoea markedly reduces anaerobe population

4. Vagina

Lactobacilli are the dominant species in childbearing years, creating by fermentative activity a pH of less than 4.5 as a consequence of oestrogen stimulation of glycogen production by vaginal epithelium

Anaerobes: Anaerobic streptococci, *Bacteroides* species

Streptococci including *Strep. agalactiae* (gp B streptococci) in 30% of normal adults

Staphylococcus epidermidis (dominant aerobe before puberty)

Staphylococcus aureus is present in about 5% of adult women

Diphtheroids

Gardnerella vaginalis is part of the normal flora but together with anaerobes becomes predominant in non-specific vaginosis

Enterobacteriaciae are present in small numbers, may become predominant flora after menopause

Mycoplasma hominis and *Ureaplasma urealyticum*, especially in sexually active women

5. Skin
Factors affecting colonisation include:

a. High humidity ⎫   increase number of bacteria
   Skin occlusion ⎭

b. High pH, soaps   increases number of bacteria

c. Sebaceous secretions   are split by bacterial lipases, releasing
   (after puberty)   free fatty acids and reducing pH. Inhibits
                          colonisation by some pathogens (e.g.
                          Streptococcus pyogenes,
                          dermatophytes)

d. Skin disease   results in colonisation with
                          Staphylococcus aureus

e. Antibiotics   promote overgrowth of yeasts, especially
                          in moist areas.

Bacteria live on the skin and its appendages in microcolonies, rather than uniformly distributed.

Normal flora consists mostly of Gram-positive organisms, including lipophilic species, as follows:

Propionibacterium acnes: The dominant flora of adults. Lipolytic
Staphylococcus epidermidis
Staphylococcus aureus
Micrococci
Diptheroids
Bacillus species
Streptococci: *Viridans* as well as anaerobic streptococci
Gram-negative rods: Small numbers of coliforms and
    pseudomonads in skin folds
Mycobacteria: *M. smegmatis, gordonae, terrae.*
Lipophilic yeasts (*Pityrosporum ovale* on scalp, *P.orbiculare* on chest). Non-pathogenic
Candida species detected on skin of 0–10% of adults, increases with old age
Dermatophytes may be present in the absence of tinea

Antimicrobial agents 1

Substances, both synthetic and naturally occuring which inhibit the growth of microorganisms in high dilution (excludes antiseptics and disinfectants).

Antibiotics, as defined by Waksman (1942) are substances produced by microorganisms, antagonistic to the growth of others. The term is now often used to describe all systemically administered antimicrobials.

The term antibiotic also encompasses some compounds used primarily as antitumour agents.

Major classes of antimicrobial agents and their spectrum of activity (see also Table 66)

A. Synthetic

1. Sulphonamides	Bacteria, chlamydia, rickettsiae and some protozoa
2. Diaminopyrimidines	As above
3. Imidazoles	Bacteria, fungi, protozoa, metazoa
4. Isonicotinic acid derivatives	Mycobacteria
5. Base analogues	Viruses
6. Quinolines	Protozoa (malaria)
7. Nitrofurans	Bacteria
8. Quinolones	Bacteria

B. Naturally occuring

1. Beta lactams, (penicillins and cephalosporins)	Bacteria (excluding mycoplasma)
2. Aminoglycosides	Aerobic bacteria
3. Chloramphenicol	Bacteria (including chlamydia, rickettsiae and mycoplasmas)
4. Tetracyclines	
5. Macrolides	Bacteria
6. Polyenes	Fungi

131

Table 66 Development of the major classes of systemic antimicrobials (in order of introduction to clinical medicine)

Year	Compound	Class	Origin
1619	Cinchona bark	Alkaloids	Plants
1906	Salvarsan	Arsenicals	Synthetic
1935	Prontosil	Sulphonamides	Synthetic analogues of
1941	Dapsone	Sulphones	para-aminobezoic acid
1940	Penicillin	Penicillins	*Penicillium* species
1944	Streptomycin	Aminoglycosides	*Streptomyces* species *Micromonospora* species
1944	Nitrofurantoin	Nitrofurans	Synthetic
1947	Chloramphenicol	Chloramphenicol	*Streptomyces venezualae*
1948	Tetracycline	Tetracyclines	*Streptomyces* species
1948	Cephalosporin C	Cephalosporins	*Cephalosporium* species *Streptomyces* species
1950	Nystatin	Polyenes	*Streptomyces* species
1952	Erythromycin	Macrolides	*Streptomyces* species
1952	Isoniazid	Nicotinic acid analogues	Synthetic
1953	Pyrimethamine	Diaminopyrimidines	Synthetic
1956	Vancomycin	Vancomycin	*Streptomyces orientalis*
1958	Griseofulvin	Griseofulvin	*Penicillium griseofulvum*
1961	Cycloserine	D-alanine antagonist	*Streptomyces orchidaceus*
1962	Lincomycin	Lincosamines	*Streptomyces lincolnensis*
1962	Nalidixic acid	Quinolones	Synthetic
1963	Metronidazole	Nitroimidazoles	Synthetic
1965	Rifamycin B	Rifamycins	*Streptomyces mediterranei*
1965	Idoxuridine	Nucleoside analogues	Synthetic
1965	Fusidic acid	Steroids	*Fusidium* sp and others
1969	Clotrimazole	Tritylimidazoles	Synthetic
1971	Clofazimine	Phenazine dye	Synthetic

Definitions

1. The minimal inhibitory concentration (MIC)
   — is the lowest concentration of an antimicrobial agent which inhibits growth of an organism
2. The minimal bactericidal concentration (MBC)
   — is the lowest concentration of an antimicrobial agent which kills the organism
3. Synergy
   — describes the ability of two antibiotics to inhibit or destroy an organism at concentrations lower than either could achieve singly
4. Antagonism
   — is said to occur if a combination of two antibiotics is less inhibitory or destructive than either would be alone
   — is usually observed with combinations of bactericidal and bacteristatic drugs
5. Bactericidal antibiotics
   — destroy bacteria (over a period of hours)
6. Bacteristatic antibiotics
   — prevent the multiplication of bacteria

Sites of action of antimicrobial agents

1. DNA replication
Nalidixic acid— inhibits DNA gyrase
Nitroimidazoles— are reduced by anaerobes to compounds which bind to DNA causing strand breakages
Novobiocin— inhibits DNA gyrase
Nitrofurans— are reduced to compounds which cause DNA strand breakages.
Nucleoside analogues— inhibit DNA polymerase or are incorporated into DNA (preventing transcription by mRNA)

2. RNA synthesis
Rifamycins— inhibit RNA polymerase
Ribavirin— inhibits viral RNA polymerase

3. Folic acid synthesis
a. Structural and functional analogues of para-aminobenzoic acid (PABA):
   Sulphonamides
   Sulphones
   Para-aminosalicylic acid (PAS)
b. Folate reductase inhibitors:
   Diaminopyrimidines

4. Cell wall (bacterial peptidoglycan)
Betalactams (penicillins and cephalosporins) inhibit transpeptidases and other enzymes
Cycloserine prevents conversion of L-alanine to D-alanine
Vancomycin, Bacitracin inhibit lipid carriage of sugar units to growing mucopeptide molecule

5. Cell membrane
Compounds which disrupt the cell membrane are usually complex circular molecules, attach to cell wall lipoprotein, increasing cell permeabilty. Tend to be toxic to mammalian cells also.
Polypeptides
Polyenes
Novobiocin

6. Protein synthesis
a. *Puromycin* — is a structural analogue of the terminal adenosine of tRNA. After incorporation into the growing polypeptide, inhibits further peptide bond formation (the compound is used experimentally, not clinically).
b. Binding to 30 S subunit of the ribosome:
   Aminoglycosides bind to 30S subunit of ribosome, distort codon-anticodon interaction, cause misreading and faulty protein synthesis
   Tetracyclines enter prokaryotic cells with the aid of a permease, bind to 30S subunit, inhibit aminoacyltRNA binding
c. Compounds which bind to the 50S subunit of the ribosome
   — do not bind to eukaryotic ribosomes
   — may be rendered ineffective by bacterial mutation
   — tend to be bacteristatic
   Chloramphenicol inhibits peptidyltransferase and peptide bond formation
   Macrolides, Lincosamines, Fusidic acid inhibit translocation

Tests of antimicrobial sensitivity

1. Dilution methods
— the antibiotic is incorporated into agar or added to broth solutions which are seeded with the organism

2. Disc diffusion
— agar plates are seeded with suspensions of bacteria on which are placed paper discs containing antibiotics
— after growth has occurred, the zone of inhibition around the disc is recorded
— Zone diameter of inhibition is inversely related to minimal inhibitory concentration

Factors affecting result of sensitivity tests
1. Size of bacterial inoculum
2. Growth conditions
   — pH, electrolyte content, tonicity
   — Presence of protein, albumin (binds antibiotic)
   — presence of carbohydrate (changes pH if organism is fermentative)
   — gas phase
   — temperature
   — presence of antibacterial antagonists, e.g.:
       (i) para-aminobenzoic acid — sulphonamides
       (ii) alanine — cycloserine
       (iii) cytosine — flucytosine
   — antibiotic stability
   — growth phase of organism
Additional factors affecting zone diameter when the test is performed on agar:
1. Antibiotic content of disc
2. Rate of growth
3. Thickness of agar
4. Accuracy of measurement of zone diameter

Development of antibiotic resistance
Populations of resistant organisms appear as a result of:
1. Selection
   — of naturally resistant bacteria from mixed populations which include sensitive and resistant species
2. Mutation
   — of some species in the presence of certain antibiotics (see Table 67).

Table 67

Organism	Agents	Clinical implications
Mycobacterium tuberculosis	Antituberculous drugs	Tuberculosis must be treated with at least two drugs to which the organism is sensitive
Neisseria	Penicillin	Gradual reduction in susceptibility resulting from mutation over the last 40 years has necessitated larger doses for single dose treatment.
Staphylococcus	Erythromycin Rifampicin Fusidic acid	Makes single use of these compounds difficult without incurring resistance
Pseudomonas aeruginosa	Aminoglycosides	Slight reduction in susceptiblity makes therapy difficult

3. Gene transfer
   — is mediated mostly by plasmid transfer in Gram-negative
     bacteria (see chapter on Bacterial Classification)
   — accounts for most resistance amongst coliform bacteria

Mechanisms of antibiotic resistance

1. Inactivation of antibiotic by enzymes, e.g.

Penicillins   ⎱
Cephalosporins ⎰ beta lactamases

Aminoglycosides: phosphorylation, acetylation, adenylation

2. Alteration of target sites for antibiotic action, e.g.

Chloramphenicol ⎫
Tetracyclines   ⎬ Alteration in 50S component of ribosome
Erythromycin   ⎭

Penicillins: Alteration in penicillin-binding proteins

3. Development of cell wall impermeability to antibiotic
4. Development of altered metabolic pathways

Antibiotic prescribing policies

— are often introduced in hospitals or regions for the following
  reasons:
  1. To prevent the emergence of antibiotic resistance
  2. To reduce the cost of antibiotic use
  3. To prevent antibiotic toxicity

Adverse effects of antibiotics:

1. Superinfection
e.g.

Gram-negative rods	colonise the pharynx, skin, female genital tract. May cause invasive infection, septicaemia
Candida species	colonise mucous membranes causing thrush, intertrigo
Clostridium difficile	multiplies and produces toxin in colon (antibiotic-associated colitis) after elimination of normal anaerobic flora

2. Hypersensitivity
Antibiotics or their metabolites may act as haptens, binding to host
protein and becoming immunogenic. The immunological
mechanisms which result in disease are:

I IgE antibody	Anaphylaxis
II Cytotoxic antibody	Haemolysis, thrombocytopenia, neutropenia
III Immune complexes	Serum sickness
IV Delayed hypersensitivity	Maculopapular skin eruptions Fever

3. Toxicity
Antibiotics may be inherently toxic (see Table 68)

Table 68 Adverse effects of antibiotics

Class	Antibiotic	Mechanism
RENAL		
Aminoglycosides	All, to a variable degree	Accumulation in proximal tubular cells.
Betalactams	Methicillin, occasionally others	Interstitial nephritis
	Cephaloridine	Accumulation in proximal tululular cells
Polyenes	Amphotericin	Membrane damage
Polypeptides	Polymyxin	Membrane damage
Tetracyclines	Older compounds	Accumulation causes increased catabolism and azotaemia
Sulphonamides	Old compounds (e.g. sulphadiazine)	Crystal deposition in tubules and ureters
Base analogues	Base analogues (flucytosine, acyclovir)	Crystal deposition
HEPATIC		
Sulphonamides	All (rare)	Cholestatic jaundice
Tetracyclines	All	Acute fatty liver, in late pregnancy
Penicillins	Isoxazoles	Jaundice
Erythromycin	estolate	Cholestatic jaundice
Fusidic acid	Fusidic acid	Jaundice
Imidazoles	Ketoconazole	Jaundice
Rifamycins	Rifampicin	Jaundice
Isoniazid	Isoniazid	Hepatitis
NEUROLOGICAL		
Tetracyclines	? All	Benign intracranial hypertension
Chloramphenicol	Chloramphenicol	Optic neuritis
Betalactams	Those that accumulate in renal failure	Encephalopathy

Class	Antibiotic	Mechanism
Aminoglycosides	All	1. Ototoxicity 2. Neuromuscular blockade
Nitrofurans	Nitrofurantoin	Peripheral neuritis
Quinolones	Nalidixic acid Oxolonic acid	Visual disturbances, Hallucinations, Seizures
		Benign intracranial hypertension
Isoniazid	Isoniazid	Peripheral neuritis (prevented by pyridoxine)
Ethambutol	Ethambutol	Optic neuritis
Base analogues	Vidarabine Acyclovir	Tremor, organic brain syndrome
Imidazoles	Metronidazole	Ataxia, peripheral neuropathy
Nitrofurans	Nitrofurantoin	Peripheral neuropathy
Vancomycin	Vancomycin	Ototoxicity
Cycloserine	Cycloserine	Seizures
PULMONARY		
Nitrofurans	Nitrofurantoin	Lung infiltrate
HAEMATOLOGICAL		
Sulphonamides	All	Agranulocytosis Haemolytic anaemia
Diaminopyrimidines	Pyrimethamine (occasionally trimethoprim)	Folate deficiency
Chloramphenicol	Chloramphenicol	Aplastic anaemia 1. dose-related 2. idiosyncratic
Rifamycins	Rifampicin	Thrombocytopenia
Betalactams	All	Neutropenia
	Carbenicillin Ticarcillin	Prevention of platelet aggregation
	Latamoxef Cephamandole	Factor V antagonist
Polyenes	Amphotericin	Thrombocytopenia

Class	Antibiotic	Mechanism
Base analogues	Flucytosine Vidarabine Acyclovir	Agranulocytosis
Vancomycin	Vancomycin	Neutropenia
GASTROINTESTINAL		
Aminoglycosides	Neomycin	Malabsorption

Antibacterial agents 2:

Major classes of antibacterial agents

SULPHONAMIDES

— Synthetic compounds which prevent the conversation of paraaminobenzoic acid to folic acid (such a step does not occur in mammals)
— Variable rates of absorption, protein binding and excretion
— Spectrum of antimicrobial activity is very broad but resistance is widespread
— The numerous members of this class differ only in pharmacological characteristics (Table 69)

Table 69

	Pharmacology	Indications
Sulphafurazole	Rapidly excreted	Urinary tract infections
Sulphadimidine		
Sulphamethoxazole Sulphamethoxydiazine	Slowly excreted	Used with trimethoprim
Sulphadoxine	Very slowly excreted	Used with pyrimethamine
Phthalylsulphathiazole	Nonabsorbed	Bowel infections and bowel 'sterilisation'
Sulphasalazine	Nonabsorbed	Treatment of ulcerative colitis

— Most important uses are:
   a. urinary tract infection
   b. some exotic infections: *Nocardia, Pneumocystis, Chlamydia*
— Side effects include:
   a. *Hypersensitivity*
      fever, rash, Stevens–Johnson syndrome
      cholestatic jaundice
   b. crystal deposition in renal tubules

c. agranulocytosis
d. haemolytic anaemia (in patients with G6PD deficiency)
e. kernicterus

DIAMINOPYRIMIDINES

— Synthetic compounds which inhibit folate reductase thereby preventing the conversion of folic to folinic acid
— Synergistic with sulphonamides
— Broad spectrum of antibacterial activity
— Compounds include:
  1. *Trimethoprim* inhibits bacterial enzyme. Combined with sulphamethoxazole (cotrimoxazole, trimethoprim compund
  2. *Pyrimethamine* inhibits protozoan enzyme. Combined with sulphadoxine (Fansidar®) or dapsone (Maloprim®)
  3. *Methotrexate* inhibits mammalian enzyme. Anticancer agent
— Uses include:
Trimethoprim
  a. urinary tract infections
  b. respiratory tract infections
  c. exotic infections: typhoid fever, brucellosis, nocardiosis
Pyrimethamine
  a. malaria
  b. toxoplasmosis
— Side effects include:
  a. folate deficiency — occurs much more frequently with pyrimethamine than trimethoprim. Reversed by folinic acid
  b. teratogenesis

TETRACYCLINES

— A family of closely related antibiotics with four-ringed structure
— Absorbed by mouth although in the case of the older compounds, much diminished by chelation with divalent cations such as Ca^{++} (milk, antacids) and Fe^{++}
— The various tetracycline compounds have much the same spectrum of antibacterial activity but differ pharmacologically, viz.

Tetracycline ⎫	Moderately well absorbed
Chlortetracycline ⎬	Excreted in bile and urine
Oxytetracycline ⎭	
Demethylchlortetracycline	More slowly excreted
Rolitetracycline	Highly water soluble, used parenterally
Methacycline ⎫	
Minocycline ⎬	Well absorbed, more lipid soluble,
Doxycycline ⎭	highly protein bound, more slowly excreted

— Broad spectrum of antibacterial activity which includes anaerobes, mycoplasmas, chlamydiae and rickettsiae. Resistance amongst coliforms is widespread
— Important uses include:
1. Respiratory tract infections (*Pneumococcus, Haemophilus*)
2. Mycoplasma pneumonia
3. Non-gonococcal urethritis
4. Acne vulgaris
5. Rickettsial infections: typhus, spotted fevers, Q fever
6. Chlamydial infections: trachoma, lymphogranuloma venereum
— Side effects:
1. Superinfection
2. Nausea, vomiting, diarrhoea
3. Staining of calcifying teeth (avoid tetracyclines in pregnancy and early childhood)
4. Aggravation of symptoms of uraemia in patients with chronic renal failure (drug accumulates, promotes protein breakdown)
5. Fatty degeneration of liver (in pregnancy)
6. Benign intracranial hypertension

MACROLIDES

— Antibiotics consisting of a macrocyclic lactone ring to which sugars are attached
— Includes erythromycin, oleandomycin, spiramycin
— Erythromycin is excreted in both urine and bile, Highly protein bound
— Erythromycin is acid labile. For oral use it is prepared as:
1. base – formulated in enteric-coated capsules
2. stearate — hydrolysed in the gut to release the base
3. propionyl ester — hydrolysed in the circulation to release the base
— Major activity is against Gram-positive rather than Gram-negative organisms but also active against mycoplasmas, chlamydiae, anaerobes and *Legionella pneumophila*
— Important uses include:
1. Respiratory tract infections
2. *Mycoplasma* pneumonia
3. Pertussis. Eliminates the organism but does not ameliorate symptoms
4. Diphtheria. Eliminates the pathogen but does not ameliorate the symptoms
5. Pneumonia due to *Legionella pneumophila*
6. Chlamydial infections: urethritis, neonatal pneumonia
7. Bowel 'sterilisation' (reduces numbers of anaerobes in colonic lumen)

— Side effects:
1. Nausea
2. Cholestatic jaundice (only with erythromycin estolate)

CHLORAMPHENICOL

— An antibiotic of relatively simple molecular structure now manufactured synthetically at low cost
— Broad spectrum of antibacterial activity although resistance amongst Gram-negative bacteria is widespread
— Because of poor solubility, chloramphenicol is administered as esters which are hydrolysed either in the gut or the circulation to yield the biologically active compound, viz.
Succinate — intravenous — hydrolysed by liver at highly variable rates in different individuals
Palmitate — oral — hydrolysed in the gut lumen before absorption
— Blood levels achieved by the oral palmitate exceed those of the intravenous succinate
— Widely distributed in tissues and body fluids, including cerebrospinal fluid
— Conjugated in the liver to the glucuronide before excretion via kidney (the glucuronide has no antibacterial activity)
— Major uses:
Because of the risk of marrow aplasia, in countries which can afford more expensive antibiotics, chloramphenicol is only used for certain life threatening infections, viz.
a. Meningitis, especially due to *Haemophilus influenzae*
b. Typhoid fever
c. infections due to betalactamase-producing *Haemophilus influenzae*
— Side effects:
a. Candidiasis
b. Bone marrow aplasia, which occurs in two forms:
Dose-related
— occurs when drug level exceeds 25 mg/l (about 50 mg/kg/d) for more than a few days.
— Bone marrow biopsy reveals cytoplasmic vacuolation of erythroid precursors
— Recovers when drug is ceased
Idiosyncratic
— Begins weeks or months after treatment and is unrelated to dose, duration or route of administration
— Occurs about once in every 40 000 therapeutic courses, often fatal
c. 'Grey syndrome' in neonates
— due to the inability of premature neonates to conjugate and excrete the drug
d. Optic neuritis

BETALACTAMS

— A large group of antibiotics characterised by the presence of a betalactam ring. Many thousands of semisynthetic compounds have been prepared
— Attach to penicillin binding proteins (PBPs) which are transpeptidases situated on the cytoplasmic membrane and responsible for peptidoglycan synthesis

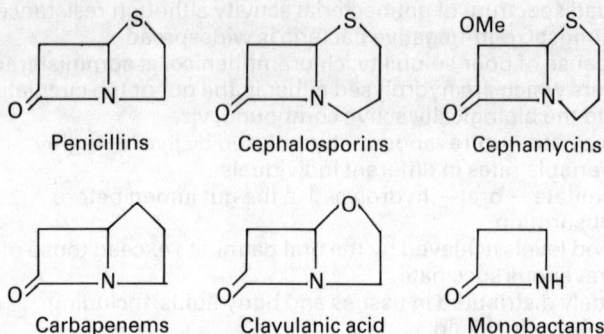

Fig. 1 Ring structures of betalactams

Table 70

Class	Origin	Ring structure
1. Penicillins	Penicillium	Penicillanic acid
2. Cephalosporins	Cephalosporium	7 aminocephalosporanic acid
3. Cephamycins	Streptomyces	7-methoxycephalosporins
4. Carbapenems	Streptomyces spp	5-membered ring is saturated
5. Clavulanic acid	Streptomyces clavuligerus	Oxygen replaces sulphur of penicillins
6. Monobactams	Chromobacterium violaceum	Single ring structure

Pharmacological characteristics of betalactams:
Unlike compounds within many other classes of antimicrobial agents, betalactams exhibit great diversity in absorption, distribution and excretion

1. Absorption
Many betalactams are not stable at the pH of gastric juice and must be given parenterally.

2. *Distribution*
Degree of protein binding ranges from 10–20% to greater than 95%, much influencing levels in tissue fluids and rate of excretion.
   Levels in CSF are generally only 1% of plasma levels but improve in the presence of inflamation.

3. *Excretion*
a. Renal
   All penicillins are excreted by glomerular filtration and most by the 'anion pump' of the proximal tubule as well (Exceptions are cephaloridine, cefoperazone and latamoxef). Half lives tend to be short.
      Probenecid blocks renal tubular secretion and prolongs half lives.
b. Metabolism
   Some (cephalothin, cefotaxime) are hydrolysed to less active desacetyl derivatives.
c. Biliary
   Many betalactams are excreted in bile.

Betalactams in current use:

1. *Penicillins*
a. Simple penicillins
   Benzyl penicillin was the first (naturally occuring) compound introduced. Still the parenteral antibiotic of choice for serious streptococcal, pneumococcal and meningococcal infections
   Procaine penicillin   } Less soluble forms of benzyl penicillin,
   Benzathine penicillin } more slowly excreted
b. Acid-resistant penicillins
   Phenoxymethyl penicillin] Incompletely absorbed but useful for
   Phenethicillin   } Gram-positive infections, especially
   Propicillin   ] streptococcal. No activity against
                                              Gram-negatives (*N. gonorrhoeae*,
                                              Bacteroides)
c. Amino penicillins
   These compounds extend the spectrum of benzyl penicillin to include *Haemophilus influenzae* and *Esch. coli* but acquisition of plasmid-determined betalactamases by these species has diminished efficacy in recent years.
   Ampicillin was the first compound of this class
   Amoxycillin is better absorbed than ampicillin
   Pivampicillin: better absorbed than ampicillin but hydrolysed in blood and tissues to form ampicillin
   5–10% of patients who receive aminopenicillins develope a characteristic measles-like maculopapular rash which may appear some days after the drug has been ceased. Allopurinol and glandular fever increase the incidence of this phenomenon which does not seem to be due to hypersensitivity.

d. Betalactamase-resistant penicillins
Penicillins which are resistant to the betalactamase of
Staphylococcus aureus. No useful activity against Gram-
negatives
Methicillin was the first compound introduced. Occasionally
causes interstitial nephritis
Cloxacillin ⎫
Flucloxacillin ⎟ Isoxazole penicillins. Partially absorbed by mouth.
Oxacillin   ⎬ Highly protein bound
Dicloxacillin ⎭
Nafcillin has had most use in the USA.

e. Carboxypenicillins
Active against *Pseudomonas aeruginosa*, indole-positive
Proteus. Resistant to degradation by betalactamase of
Bacteroides fragilis and indole-positive *Proteus* but not those of
Staphylococcus aureus or many coliforms. Large doses must be
used (12–30 g/day).
Carbenicillin was the first introduced.
Ticarcillin is 2–4 times more active than carbenicillin
Indanylcarbenicillin: orally-absorbed derivative of carbenicillin
useful for renal tract infection due to *Ps. aeruginosa*

f. N-acyl derivatives of ampicillin
Spectrum of activity is similar to carboxypenicillins but with more
activity against *Klebsiella aerogenes*.
Azlocillin Mezlocillin Ureido penicillins
Piperacillin

g. Amidino penicillins
These compounds lack activity against Gram-positives and
Haemophilus influenzae but have good activity against
ampicillin-resistant *Esch.coli*, *Klebsiella* and *Enterobacter*. Bind
only to PBP2, creating osmotically stable round forms.
Mecillinam is the only representative of the group, known as
amdinocillin in the USA.

2. *Cephalosporins*
a. 'First generation'
— Compounds with good activity against *Staph. aureus* and
some Enterobacteriaceae (*Esch.coli*, *Klebsiella*, *Proteus
mirabilis*)
— Less active than ampicillin against *H. influenzae*
— Choice of compound is based on pharmacological attributes,
viz.
Cephalothin: Rapidly excreted
Cephaloridine: Slowly excreted (not secreted by renal tubule).
Nephrotoxic in doses greater than 4 g/day
Cephalexin ⎫ Well absorbed by mouth
Cephradine ⎭
Cefazolin: Slowly excreted, less painful intramuscularly

b. 'Second generation' cephalosporins
— Greater spectrum of activity against coliforms, good activity against *Haemophilus influenzae*, including betalactamase-producing strains
Cefamandole is only relatively resistant to degradation by Gram-negative betalactamases
Cefuroxime has had most use in the UK
Cefoxitin is the first cephalosporin with reasonable activity against *Bacteroides fragilis*

c. 'Third generation' cephalosporins
— Considerable resistance to degradation by betalactamases of most Enterobacteriacae which are usually inhibited by 1 mg/1 of these compounds. Some have antipseudomonal activity
— Some species (*Enterobacter*, *Serratia* and *Pseudomonas*) contain inducible betalactamases which may cause the development of resistance during therapy
— Activity against *Staphylococcus aureus* is significantly less than the older compounds
— Levels in cerebrospinal fluid are sufficient to treat gram-negative meningitis
Cefotaxime was the first of this group developed
Cefoperazone ⎱ contain a methylthiotetrazole side chain
Latamoxef ⎰ associated with prolonged prothrombin time
Ceftriaxone ⎱ Newer compounds with slower rates of
Ceftazidime ⎰ excretion and greater antipseudomonal activity

3. Cephamycins

Cefoxitin, the only member of this class of betalactam, is usually described as a cephalosporin (see 'second generation' cephalosporins, above)

4. Carbapenems

These compounds contain neither sulphur nor oxygen in the ring fused to the betalactam ring.

Chemical stability has been achieved by creating an N-formimidoyle derivative but the compound is metabolised by a dehydropeptidase present in the basement membrane of the proximal renal tubule.

Very broad spectrum of antibacterial activity as well as acting as betalactamase inhibitors.

The agent currently undergoing trial of clinical efficacy consists of:
a. N-Formimidoyl thienamycin (imipenem), together with
b. an inhibitor of renal dehydropeptidase I

5. Betalactamase inhibitors

Betalactams which may not have antibacterial activity but which inhibit betalactamases. Can be combined therapeutically with other betalactams.

Clavulanic acid was the the first compound introduced. Inhibits betalactamases of *Staphylococcus aureus* and many Gram-negatives. Combined with ampicillin (Augmentin®)
Thienamycin: A potent betalactam also capable of inhibiting betalactamases
Sulbactam: Combined with ampicillin (Sultamicillin®)

6. Monobactams

Single-ringed betalactams with excellent activity against Gram-negative bacteria but no activity against Gram-positive bacteria or anaerobes.

Azthreonam, the first of these compounds, is currently undergoing trail.

Side effects of betalactams

1. Hypersensitivity

Betalactams or their metabolites may act as haptens by binding to tissue proteins and stimulating the immune system.

About 1–8% of courses of treatment with betalactams are estimated to result in some type of allergic reaction.

Agricultural, industrial and medical contamination of the human environment causes occult sensitisation, so that 70–90% of adults have detectable levels of penicillin antibodies.

The determinants of allergy to benzylpenicillin are as follows:
Penicilloyl: The 'major' determinant, accounts for 95% of the tissue bound compounds.
Penicilloyl polylysine is used as a skin test for immediate hypersensitivity
Penicillin
Penicilloate } 'Minor' determinants, accounting for 5% of bound
Penilloate } compounds

Mechanisms and manifestations of penicillin hypersensitivity include:

a. Anaphylaxis (IgE)
  — Urticaria, laryngeal oedema, bronchospasm
  — Occurs in about 0.2% of treatments, fatal in 0.001%
b. Cytotoxic antibodies
  — Haemolysis, neutropenia, thrombocytopenia following attachment of penicilloyl to cell membranes during high-dose therapy
c. Immune complexes
  — Serum sickness (arthralgia, nephritis, etc)
d. Others
  — Maculopapular rash
  — Erythema nodosum, erythema multiforme
  — Fever
  — Eosinophilia
  — Interstitial nephritis

2. *Thrombophlebitis*
— at sites of intravenous administration

3. *Encephalopathy*
— follows high-dose therapy in patients with impaired renal function

4. *Bleeding*
— has three mechanisms:
a. *Thrombocytopenia* due to hypersensitivity. May follow the use of any betalactam
b. *Diminished platelet aggregation* may follow use of betalactams in high doses (benzyl penicillin, carbenicillin, ticarcillin)
c. *Inhibition of vitamin K-related coagulation factors*: Reported with latamoxef, cephamandole

5. *Sodium overload*
— especially follows use of disodium salts (eg, carbenicillin, ticarcillin)
— may aggravate hypokalaemia

6. *Neutropenia (see hypersensitivity)*

7. *Nephrotoxicity*
— has two causes:
a. *Interstitial nephritis* occurs most commonly with methicillin. Fever, rash, eosinophilia, microscopic haematuria.
b. *Proximal tubular necrosis* occurs almost exclusively with cephaloridine which accumulates in proximal tubular cells

Mechanisms of resistance to betalactams
1. *Betalactamases* account for most resistance In *Staph. aureus* and Enterobacteriaciae
2. *Decreased affinity for penicillin binding proteins (PBPs)*
   Multiresistant S. aureus (MRSA)
   Streptococcus faecalis
   Penicillin-insensitive strains of pneumococcus, gonococcus
3. *Cell wall impermeability* accounts for the resistance of Gram-negatives to earlier betalactams as well as the continued resistance of *Pseudomonas aeruginosa*

Betalactamases
— are enzymes which hydrolyse the betalactam ring of these antibiotics, destroying their antibacterial activity
— includes a great range of enzymes classified according to substrate profiles, isoelectric point and the compounds which inhibit them
— produced by many bacteria but with characteristic differences between those produced by Gram-negative and Gram-positive species (Table 71)

Table 71   Characteristics of betalactamases

	Gram-positive	Gram-negative
Site	Diffuse into the surrounding medium	Confined to periplasmic space.
Amount produced	Considerable	Relatively little
Types	Few	Many
Method of production	Mostly inducible (presence of drug increases enzyme synthesis)	Mostly constitutive

AMINOGLYCOSIDES

Glycosidic derivatives of streptamine.
   Most pharmacological attributes are shared by the group, viz.
1. Not absorbed by mouth
2. Minimal protein binding
3. Occupy extracellular space, little penetration of cerebrospinal fluid
4. Excreted only by glomerular filtration. Accumulate rapidly in the presence of renal failure
5. Following glomerular filtration, a proportion of these drugs are reabsorbed into proximal tubular cells, causing nephrotoxicity
6. All are potentially neurotoxic (VIIIth nerve)
7. Primarily active against aerobic gram negative rods. No activity against anaerobes and poor activity against gram positive organisms
8. Determinations of circulating levels of aminoglycosides are important because
   a. the doses usually prescribed are calculated to barely exceed the MIC of most Gram-negative pathogens
   b. these potentially neurotoxic drugs accumulate readily in the presence of renal dysfunction and are themselves a cause of renal failure
   Resistance, if it develops, may be due to
1. *Mutation*, causing alteration of ribosomes or cell wall impermeability
2. *Acquisition of plasmids* which code for the production of inactivating enzymes
   Compounds in current use:

1. Topical (too toxic for systemic use)
Neomycin
Framycetin (Soframycin®)

2. Systemic (Table 72)

Table 72 Aminoglycosides for systemic use

	Origin	Semisynthetic derivatives
Streptomycin	*Streptomyces griseus*	
Kanamycin	*Streptomyces kanamyceticus*	Amikacin
Gentamicin	*Micromonospora purpurea*	
Sisomicin	*Micromonospora inyoensis*	Netilmicin
Tobramycin	*Streptomyces tenebrarius*	

Major uses
1. Sepsis due to aerobic gram negative rods
   e.g. renal tract infection, peritonitis, cholangitis
2. Tuberculosis
   Streptomycin is bactericidal for *M. tuberculosis* and is often used in the initial treatment
3. Topical
   Adverse effects of topically applied aminoglycosides include the development of resistance, hypersensitivity and systemic toxicity following application to large areas of denuded skin
   a. Skin, nasal vestibule, external ear
   b. Conjunctiva, subconjunctiva
   c. Bowel 'sterilisation', treatment of *Esch.coli* enterocolitis.
   d. Intrathecal, intraventricular
   e. Bladder irrigation
   f. Lower respiratory tract (aerosol)

Side effects

1. Ototoxicity
Aminoglycosides all have the potential to damage the eighth nerve as follows:
Cochlear division: Results in deafness, usually irreversible
                        More common with kanamycin, amikacin
Vestibular division: Results in ataxia
                        More common with streptomycin, gentamicin

2. Nephrotoxicity
— occurs in 10–25% of patients given gentamicin, somewhat less with tobramycin, amikacin and netilmicin
  developes after 3–5 days, may continue for 7–10 days after the drug is ceased
— reversible in 30–60 days
— may be potentiated by other nephrotoxic agents such as hypotension, cis-platinum, cephalothin and frusemide

3. Hypersensitivity

4. Neuromuscular blockade

LINCOSAMINES

— Antibiotics derived from *Streptomyces lincolnensis*
— Well absorbed by mouth and widely distributed through total body water
— Although there is no chemical relationship, antibacterial activity and mode of action (on bacterial ribosome) is similar to erythromycin
— Most active against gram positive cocci and anaerobes
— Two compounds are in clinical use:
  1. Lincomycin
  2. Clindamycin — a semisynthetic derivative of lincomycin, better absorbed and much more active

Major uses:
1. Staphylococcal infections, especially when betalactams are contraindicated
2. Anaerobic infections

Side effects:
1. Hypotension may follow rapid intravenous infusion
2. Antibiotic-associated colitis is relatively common after use of this group of antibiotics

IMIDAZOLES

A diverse group of synthetic compounds having in common only the presence of an imidazole ring.
  Excreted in urine as well as being metabolised.

1. Antianaerobic
These imidazoles are reduced by strict anaerobes to compounds which bind to DNA causing strand breakages. Well absorbed by mouth. Spectrum of activity includes two groups of organisms, viz.
a. *Bacteria*: *Clostridia, Bacteroides*, fusobacteria, anaerobic cocci
b. *Protozoans*: *Trichomonas, Entamoeba, Giardia*
Metronidazole was the first compound introduced
Tinidazole ⎫ not metabolised, more slowly excreted
Ornidazole ⎭

2. Antifungal
These imidazoles inhibit fungal cell membrane synthesis. Spectrum of activity includes yeasts, dermatophytes and some causes of invasive disease.

Clotrimazole
Econazole   Used topically for thrush and tinea
Miconazole is used intravenously, solubilised in polyethoxylated
castor oil. Side effects are frequent
Ketoconazole is the first antifungal imidazole to be well absorbed by
mouth

3. Antiparasitic
Various compounds with various indications for use, viz.
Thiabendazole: Roundworms (intestinal and visceral)
Mebendazole: Intestinal roundworms (poorly absorbed)
Levamisole has been used as an immunostimulant
Niridazole: Schistosoma mansoni

NITROFURANS

Synthetic compounds with broad spectrum of antibacterial activity.
Includes the following:
Nitrofurazone, used topically for infected wounds.
Furazolidone, used for enterocolitis.
Nitrofurantoin, used for renal tract infections. Given orally, is well
absorbed but excreted too rapidly to produce effective plasma
levels. Important adverse effects include peripheral neuropathy and
pulmonary infiltrates

QUINOLONES

Relatively simple synthetic compounds which inhibit bacterial DNA
gyrases (the enzymes which supercoil the strands of bacterial DNA).
Nalidixic acid has been used for many years for the treatment of
urinary tract infection. Resistance develops rapidly and side effects
are frequent
Oxolinic acid⎫
Cinoxacin　⎬ are somewhat more active
Norfloxacin is a hundred times more active than nalidixic acid and
has antipseudomonal activity
　　Side effects with the older compounds include:
1. Nausea
2. Rashes (phototoxic, urticarial, maculopapular)
3. CNS distubances (visual disturbances, hallucinations, intracranial
　 hypertension, seizures)
Nalidixic acid should not be used if renal function is abormal.

RIFAMYCINS

— A group of antibiotics derived from *Streptomyces mediterranei*
— Rifampicin is the most frequently used derivative of rifamycin
— Well absorbed by mouth, excreted mainly in bile, undergoes
　 enterohepatic recirculation

— Diffuses into tissues and intracellular fluid. Red colour stains urine, sputum and tears
— Highly active and bactericidal against Gram-positive bacteria, including mycobacteria but presence of resistant mutants requires that the rifampicin should always be used with other antimicrobials

Major uses
1. Mycobacterial infections
   a. Tuberculosis
   b. Leprosy
2. Staphylococcal infections
   a. Prosthetic endocarditis due to *Staph. epidermidis*
   b. Multiresistant *Staph. aureus* (MRSA)
3. Treatment of carriers of *Neisseria meningitidis*

Side effects
1. Hepatic dysfunction
2. Immune thrombocytopenia
3. Flu-like illness
4. Hepatic microsomal enzyme induction may hasten the excretion of concurrently administered drugs such as steroids, oral contraceptives, coumarins and tolbutamide

PEPTIDES

A large group of peptide-linked amino acid antibiotics, mostly derived from the genus Bacillus, which are generally too toxic for systemic administration.

Bacitracin
Gramicidin } are only used topically

Polymyxin is occasionally used parenterally for infections due to *Pseudomonas aeruginosa*. Nephrotoxic and neurotoxic

POLYENES

Antifungal antibiotics derived from species of *Streptomyces*. Bind to ergosterol, the principal sterol in fungal cell membranes and also, to a lesser degree, the sterols of mammalian cells.

Nystatin
Pimaricin } are used topically

Amphotericin is the only systemically administered polyene antifungal
Poorly soluble (administered with bile salts)
Accumulates on cell membranes
Toxicity includes fever, hypokalaemia, anaemia, renal tubular damage

BASE ANALOGUES

Synthetic derivatives of the purine and pyrimidine bases
Flucytosine acts on yeast-like fungi. Most important use is for
cryptococcosis. Resistance developes rapidly if used alone
Idoxuridine ⎱ Antiviral agents (see chapter on General
Vidarabine ⎰ Characteristics of Viruses)
Acyclovir

VANCOMYCIN

— Derived from *Streptomyces orientalis*
— Not absorbed by mouth, excreted only by glomerular filtration
— Highly protein bound
— Only active against gram-positive bacteria and gram-negative
  cocci. Bactericidal

Major uses:
1. Staphylococcal infections, in special situations:
   a. Multiresistant *Staph. aureus*
   b. Prosthetic endocarditis
   c. Shunt infections in anephric patients. Satisfactory levels are
      achieved by weekly doses (vancomycin is not dialysable)
2. Colitis due to *Clostridium difficile* (given orally)
3. Streptococcal endocarditis, especially when the patient is allergic
   to penicillin

Side effects
1. Thrombophlebitis
2. Nephrotoxicity
3. Neutropenia
4. Ototoxicity
5. Rash, fever

FUSIDIC ACID

— A steroid derived from *Fusidium coccineum*
— Well absorbed by mouth, highly protein bound. Excreted and
  concentrated in bile
— Active against only Gram-positive bacteria and Gram-negative
  cocci but resistant mutants appear rapidly and fusidic acid
  should usually be used in conjunction with other antimicrobials

Major uses:
Staphylococcal infections
a. Multiresistant *Staph. aureus* (MRSA)
b. Osteomyelitis

Side effects
1. Nausea
2. Jaundice

Infective endocarditis

Infection of the endocardial surface of the heart with the production of vegetations.

Almost any organism, with the possible exception of viruses, may be causative.

Pathogenesis
1. Haemodynamic change caused by valvular abnormality creates turbulence which causes jet injury of endothelium
2. Platelets and fibrin are deposited on damaged endothelium
3. Coincidental bacteraemia, especially from oral cavity, leads to adherence of bacteria to this thrombus
5. Bacterial division results in formation of vegetations and valve destruction
6. Persistent antigenaemia results in immune complex formation

Organisms

1. **Streptococci**
   — account for 50–80% of cases
   Strep. 'viridans'
   — Alpha haemolytic streptococci of the oropharynx are the most frequent causes of endocarditis
   — Includes *Strep. sanguis, Strep. mitis (mitior), Strep. mutans, Strep. salivarius, Strep. milleri (intermedius)*.
   Strep. faecalis
   — follows bladder infection in young women and the elderly
   Strep. bovis
   — is an occasional cause of endocarditis associated with colonic carcinoma
   Strep. pneumoniae
   Strep. pyogenes
   — These and other streptococci are rare but potentially destructive causes of endocarditis

2. Staphylococci

Staph. aureus
— is the most frequent cause of endocarditis on a normal valve.
  Rapid onset, high mortality
Staph. epidermidis
— follows use of valve prostheses

3. Other organisms

Candida species — mostly on prosthetic valves
Pseudomonas species — in drug addicts
Anaerobes ⎫
Q fever ⎬ Rare causes of endocarditis
Gram-negative bacilli ⎪
Neisseria gonorrhoeae ⎭

UNDERLYING HEART DISEASES

1. Rheumatic heart disease
2. Atherosclerosis
3. Congenital heart disease
4. Mitral valve prolapse
5. Valve prosthesis
6. No underlying defect in most cases of *Staphylococcus aureus* infection

CLINICAL FEATURES

1. Murmur
2. Evidence of infection, including:
   — fever — almost invariably present at some stage
   — sweats, malaise, anaemia, weight loss, myalgia
   — splenomegaly
3. Embolic phenomena (pulmonary or systemic)
4. Evidence of immune complex disease: petechiae, Osler's nodes, Roth spots, Janeway's lesions, splinter haemorrhages, arthralgia
5. Clubbing (if prolonged)
6. Mycotic aneurysms
7. Cardiac failure

LABORATORY FINDINGS

1. Bacteraemia: present in 90–95% of untreated cases
2. Evidence of infection
   — high ESR
   — anaemia
   — leucocytosis, reactive bone marrow

3. Evidence of immune complex disease:
   — low serum complement
   — microscopic haematuria, red cell casts, proteinuria
   — circulating immune complexes.
   — rheumatoid factor
4. Echocardiography
   – detects vegetations in 50–60% of cases

Causes of negative blood cultures
1. Prior administration of antibiotics
2. Right-sided disease
3. Infection lasting longer than 3 months
4. Fastidious or slow-growing organisms:
   a. pyridoxine-dependent streptococci
   b. anaerobes, CO_2-dependent organisms.
5. Fungi: may grow poorly in liquid medium
6. Obligate intracellular bacteria, especially Q fever

PRINCIPLES OF TREATMENT
1. Organism should be identified to species level and the minimal
   bactericidal concentration (MBC) for relevant antibiotics
   determined
2. Antibiotics must be bactericidal
3. Antibiotic administration must be parenteral unless adequate
   (measured) levels can be achieved with oral agents
4. Anticoagulants are contraindicated except for prosthetic
   endocarditis, in which case the prothrombin time should be
   measured frequently
5. Veins should be aseptically and expertly cannulated preferably
   using steel needles which are changed every 48 hours
6. Regular investigations should include haemoglobin, white cell
   count, creatinine, aminoglycoside concentrations
7. If there is doubt about the efficacy of treatment, the bactericidal
   activity of serum against the isolated organism should be
   periodically determined

Antibiotic regimens:

1. Streptococcus 'viridans'
Penicillin G, 2 Mu every 4–6 hours for 4 weeks. Addition of
streptomycin to penicillin can produce cure in less than 4 weeks but
it is not essential unless MBC is greater than 0.5 mg/l

2. Streptococcus faecalis
Penicillin G 3–4 Mu i.v. every 4 hours PLUS aminoglycoside
(streptomycin, gentamicin or tobramycin) in appropriate dosage, for
4–6 weeks. Penicillin alone cannot cure *Strep. faecalis* endocarditis
because it is not bactericidal for the organism

3. Staphylococcus aureus

Cloxacillin, flucloxacillin, nafcillin or cephalothin in doses of 2–4g every 4 hours for 4–6 weeks. Gentamicin is often added. It hastens the resolution of fever but has not been shown to produce greater cure rates and is frequently nephrotoxic in these patients

4. Culture-negative endocarditis

These cases are treated as for Strep. faecalis

Management of patient with penicillin allergy:
1. Known before commencement of therapy:
   a. cephalothin if MBC is sufficiently low, or
   b. vancomycin, or
   c. cautious desensitisation.
2. Occurring during therapy:
   a. add prednisone 20–30 mg per day, or
   b. change to cephalothin or vancomycin

For sensitive strains of Strep. 'viridans' it is reasonable to cease treatment and observe if 3 weeks of high dose therapy has been given.

Some special associations with certain types of endocarditis

Prosthetic valves
— Early infection is usually due to Staph. epidermidis, occasionally Candida albicans
— Late infection is due to Strep. 'viridans', Staph. aureus, Staph. epidermidis

Drug addicts
— Usually right-sided
— Staph. aureus, Candida and Pseudomonas species predominate

Intracardiac complications of endocarditis
1. Aneurysm of sinus of Valsalva
2. Pannus on prosthetic valve ring (causes incompetence)
3. Valvular incompetence
4. Aortic root abscess
5. Aortic to right atrial shunt
6. Valvular occlusion
7. Complete heart block
8. Pericarditis (due to coronary embolus or spread of aortic valve infection)

Indication for surgery
1. Haemodynamic complications: see above
2. Prolonged disease, especially with large or calcified vegetations easily demonstrated by echocardiography

3. Infection refractory to antibiotics:
   a. Difficult organisms
   b. Aortic root abscesses
   c. Prosthetic endocarditis due to staphylococci

Prevention of endocarditis
— Bacteraemia may follow surgery or instrumentation at any site
   which harbours commensal flora and antibiotics should
   therefore be given prior to such surgery
— Dental surgery is the most likely surgery to predispose to
   endocarditis but fewer than 15% of patients give a history of
   recent dental extraction
— Various regimens have been proposed for the use of prophylactic
   antibiotics (see Table 74)

PERICARDITIS

— Inflammation of the pericardial sac
— Infections involving the pericardium often involve the
   myocardium as well (myocarditis)

Causes

1. *Viral*
   a. Enteroviruses
      — Coxsackie B viruses are the usual cause of pericarditis and
         often cause myocarditis as well
      — May cause progressive cardiomyopathy or late constrictive
         pericarditis
   b. Others
      — Mumps, Epstein–Barr virus, varicella, adenoviruses,
         influenza

2. *Bacterial*
   a. Pyogenic
      — *Strep. pneumoniae, Staph. aureus, meningitidis, Strep.
         pyogenes, Actinomyces*
   b. Tuberculous
      — *M. tuberculosis*, rarely other species.
   c. Rickettsial
      — Typhus, Q fever.

3. *Fungal*
   Coccidioides, Histoplasma and others

4. *Parasitic*
   Toxoplasma, Trichinella, Trypanosoma

Clinical features
1. Fever, signs of associated infection
2. Pericardial pain and friction rub
3. Tamponade is frequently associated with pyogenic but almost never with viral pericarditis
4. Cardiac failure (due to associated cardiomyopathy)
5. Slowly developing disease due to Coxsackie B viruses, tuberculosis and other causes may present with signs of constrictive disease.

Laboratory features of pericarditis
1. Evidence of cardiac effusion
   a. Cardiomegaly
   b. Low voltage ECG, ST segment elevation
   c. Sonolucent space on echocardiogram
   d. Effusion seen on computerised tomography
2. Culture of aspirated pericardial fluid will usually establish the cause of pyogenic but rarely tuberculous or viral pericarditis
3. Other microbiological techniques, viz.
   a. Blood culture (often positive in pyogenic infections)
   b. Serology for Coxsackie viruses (usually not helpful)
   c. Tuberculin test (nearly always positive in tuberculous pericarditis)

Non-infective causes of pericarditis include
1. Myocardial infarction
2. Uraemia
3. Pericardiotomy (Dressler's syndrome)
4. Malignancy
5. Collagen-vascular disease
6. Acute rheumatic fever
7. Familial Mediterranean fever

Table 73 Recommendations for the prevention of endocarditis (proposed by the British Society for Antimicrobial Therapy, 1982)

1. Dental surgery	
General dental practice	Amoxycillin, 3 g orally 1 before the operation.
General dental practice, patient allergic to penicillin	Erythromycin stearate, 1.5 g p.o. 1–2 h before the surgery, followed by a second dose of 0.5 g 6 h later
Patients who have recently been receiving penicillin	May harbour resistant organisms. Use erythromycin (as above)
Patients having a general anaesthetic (and cannot take drugs by mouth)	Amoxycillin, 1 g i.m. in 2.5 ml of 1% lignocaine before induction followed by 0.5 g p.o. 6 h later

Table 73 Contd

Patients having anaesthetic, allergic to penicillin or having had recent penicillin	Vancomycin, 1 g by slow i.v. infusion over 20–30 min, followed by gentamicin 120 mg i.v. before induction
Special risk patients (general anaesthetic plus prosthetic valve, or previous endocarditis)	Amoxycillin (as above) plus gentamicin 120 mg i.m. before induction, OR Vancomycn plus gentamicin (as above)
2. Genitourinary surgery (cystoscopy, urethral dilatation, prostatectomy)	Amoxycillin plus gentamicin, OR Vancomycin plus gentamicin
3. Obstetric and gynaecological procedures:	Only to be used in patients with prosthetic valves As for geniturinary surgery
4. Gastrointestinal procedures	Only to be used in patients with prosthetic valves As for geniturinary surgery

Urinary tract infections

Types of genitourinary infection include:
1. Urethritis
   — Usually due to sexually transmitted infection
2. Prostatitis
   Orchitis
   Epididymo-orchitis
   — May be venereal or due to coliforms
3. Cystitis
4. Pyelonephritis
   — Usually referred to as urinary tract infection
5. Renal carbuncle
   Perinephric abscess
   — Infection in perinephric fat may be metastatic (*Staph. aureus*)
   or ascending urinary (coliform) infection

Definitions

1. *Bacilluria, bacteriuria*
   Presence of bacteria in the bladder but not necessarily causing infection.

2. *Pyuria:*
   Increased excretion rate of white cells, defined as greater than 400,000 cells per hour.
   On single specimens, pyuria is determined using one of two techniques:
   a. *counting chamber* revealing greater than 10 cells per μl
   b. *centrifuged deposit* containing more than 5–10 cells per high power field

3. *Urinary tract infection*
   Defined as the presence of bacteria plus increased white cell count in bladder.

163

Factors predisposing to urinary tract infection
1. Females are more prone to infection than males because of:
   — short urethra
   — colonisation of periurethral skin with coliforms
   —urethral trauma during intercourse
2. Obstruction
   e.g., prostatic hypertrophy, urethral stricture, stones, bladder dysfunction
3. Catheterisation
4. Vesico-ureteric reflux
5. Pregnancy
6. Diabetes mellitus (if poorly controlled)

Organisms
1. *Escherichia coli* accounts for 90% of community-acquired infection
2. *Proteus* species — frequently involve renal pelvis and parenchyma
   — predispose to stone formation (urease renders urine alkaline)
3. *Klebsiella* — common in hospitals, following catheterisation
   Enterobacter
   Serratia — often resistant to multiple antibiotics
   Citrobacter
   Pseudomonas
4. *Streptococcus faecalis*
5. *Staphylococcus saprophyticus* causes cystitis in young women
6. *Staphylococcus epidermidis* occasional cause of cystitis, usually follows catheterisation
7. *Candida* species — follows catheterisation
   — occasionally follows blood stream invasion
8. *Adenovirus* is a rare cause of haemorrhagic cystitis in children
9. *Staphylococcus aureus* is an unusual cause of cystitis. May colonise the urinary tract from the circulation after an episode of bacteraemia

Clinical features
1. Dysuria, frequency, nocturia
2. Pyuria, haematuria
3. Fever
4. Loin pain and tenderness
   — indicate renal involvement (pyelonephritis)
Diagnosis of urinary tract infection is based on the detection of bacteria and white cells in urine.

Methods of urine collection
1. Voided (midstream, clean-catch) urine
   — is the most convenient way to collect the specimen but prone to contamination with vaginal organisms
   — must be examined quantitatively (small numbers of bacteria are usually due to vaginal contamination)
   — is probably contaminated if:
      a. vaginal epithelium is present
      b. bacterial count is less than 10^4 organisms per ml
      c. culture is mixed and contains vaginal organisms (such as *Gardnerella, Lactobacilli*, group B streptococci)
2. Catheter
3. Cystoscopy
   (ureteric urine can also be collected)
4. Suprapubic bladder puncture
   — especially useful in neonates

Determination of white cell count:
a. Centrifuged deposit
b. Counting chamber
c. Leucocyte esterase test (paper strip)

Methods of quantitating bacteria in urine
a. Wire loop
b. blotting paper strips
c. flooded agar slopes (dipslides)
   — confluent growth indicates presence of $> 10^5$ organisms per ml
d. drops from calibrated pipettes
e. counting chamber
f. presence of nitrite in urine (paper strip)

Plating media for bacterial culture
a. MacConkey's medium
b. CLED (cystine lactose electrolyte-deficient) medium

Localisation of urinary tract infection
The following suggest presence of infection in the renal parenchyma as well as the bladder:
a. presence of loin pain and tenderness
b. presence of white cell casts
c. presence in urine of fibrin degradation products or brush border enzymes
d. scarring seen on intravenous pyelogram
e. diminished concentrating ability
f. presence of *Proteus* species
g. presence of the same bacterial serotype in recurrent episodes

h. presence of antibodies on bacterial surface (fluorescent antibody technique)
i. swift return of bacteria after neomycin bladder washout (Fairley technique)
j. response to single dose treatment. Most infections involving the bladder alone are eliminated by a single high dose of antibiotic
k. renal biopsy showing inflammatory change or bacterial growth
l. gallium scan showing renal isotope accumulation

Asymptomatic bacteriuria
— defined as the presence of more than 10^5 bacteria per ml in consecutive samples of voided urine in the absence of pyuria or symptoms
— in females, increases in incidence with age
— does not compromise renal function unless coexistent:
a. obstruction
b. analgesic nephropathy
c. papillary necrosis
d. diabetes mellitus (poorly controlled)
  — is associated with ascending infection in pregnancy

Sterile pyuria
— is defined as pyuria in the absence of bacilluria
— has many causes
  a. poor urine collection (i.e. vaginal contamination)
  b. urinary tract infection. Organism may fail to grow because of:
    (i) antibiotics or antiseptics in urine
    (ii) intermittent excretion of organisms in chronic disease
  c. renal tuberculosis
  d. urethritis, prostatitis
  e. perinephric abscess
  f. infections due to organisms not cultured on nutrient agar (adenovirus, schistosomiasis, leptospirosis)
  g. non-infectious disease (glomerulonephritis, interstitial nephritis, polycystic disease, analgesic nephropathy)

Urethral syndrome in females
— is defined as the presence of dysuria in the absence of bacteriuria
— has several causes
  a. Urinary tract infection with low numbers of bacteria in the frequently voided urine
  b. Chlamydial urethritis
  c. Vaginitis due to Candida, Trichomonas, N.gonorrhoeae
  d. Herpes simplex involving periurethral region
  e. No causative agent found in the majority of cases

Relapsing urinary tract infection
Infection in the renal medulla is difficult to eradicate and may
relapse intermittently over long periods, each time with the same
organism.
   Causes include:
1. Stones, anatomical abnormalities
2. Chronic bacterial prostatitis may cause recurrent ascending
   infection
3. In most cases, no cause can be found. Explanations for bacterial
   persistence in renal medullary parenchyma include:
   — destruction of complement by ammonia
   — hypertonicity of renal medulla may impair leucocyte
     adhesion, phagocytosis and polymorph migration
   — inactivation of some antibiotics by low pH
   — persistence of cell wall deficient organisms (L-forms)
     following betalactam antibiotics

Recurrent urinary tract infection
Repeated infection often with different organisms, not necessarily
involving renal parenchyma.
   Prevented by:
   — voiding after intercourse
   — frequent micturition
   — continous nightly doses of nitrofurantoin or trimethoprim

Complications of renal tract infection
1. During the first five years of life retards renal growth
2. Stone formation, especially following infection by *Proteus*
   species (producing staghorn calculi).
3. Septicaemia
   Urinary tract infection, especially following catheterisation, is the
   most common cause of septicaemia in hospitalised patients
4. Metastatic infection
   Includes septic arthritis, vertebral osteomyelitis, meningitis in
   neonates
5. Renal destruction
   — only occurs in the presence of stones, reflux, analgesic
     nephropathy and poorly controlled diabetes
6. Perinephric abscess
   — follows rupture of cortical infection through the renal capsule

Principles of management
1. Use antibiotics which are concentrated in urine
2. Duration of treatment is determined by localisation
   a. a single large dose will cure cystitis and can be used as test of
      localisation
   b. parenchymal (relapsing) infection may be eradicated by
      prolonged (6 weeks) treatment

3. Making the urine alkaline (citrate, bicarbonate) alleviates dysuria and potentiates the activity of sulphonamides
4. Investigation of the renal tract is indicated in:
   a. childhood
   b. males
   c. relapsing infections in females
5. Methods of investigation should include:
   a. intravenous pyelogram
   b. micturating cystourethrogram (to detect reflux in children)
   c. cystoscopy

Antibiotics for urinary tract infection
The compounds usually chosen are orally absorbed, active against gram negative bacilli and concentrated in urine:
Sulphonamides have been used for many years. Resistance is frequent
Cotrimoxazole improves the spectrum of sulphonamides
Trimethoprim may be used in patients allergic to sulphonamides
Aminopenicillins (ampicillin, amoxycillin)
'Augmentin': addition of clavulanic acid prevents destruction of aminopenicillins by betalactamases
Nitrofurantoin: useful for recurrent infection because resistance does not develop
Quinolones: Nalidixic acid, oxolinic acid, cinoxacin, norfloxacin
Cephalosporins: Cephalexin, cephradine
Indanylcarbenicillin: active against *Pseudomonas aeruginosa*

Hexamine (methenamine) mandelate {
— is partially hydrolysed to ammonia and formaldehyde in urine of pH > 5.5
— oral ascorbic acid or ammonium chloride may be necessary to achieve this
— has only weak activity

Intracranial infections

Clinical entities include:
1. Meningitis
2. Encephalitis
3. Brain abscess
   Parameningeal infections

MENINGITIS

Infection in the subarachnoid space following:
1. Bacteraemia or viraemia
2. Direct spread of bacteria from sites of contiguous infection

The common bacteraemic/viraemic causes are age-related as follows:

Group B streptococcus: almost exclusively in neonates

Esch. coli: mostly neonates, especially due to capsular type K1

Haemophilus influenzae ⎫ occur mostly in the first five years of life
Neisseria meningitidis ⎭

Streptococcus pneumoniae has a higher incidence under 5 and over 50 years of age

Viruses ⎰ — Peak incidence in young children, rare over 40
⎱ — Common causes include mumps, enteroviruses, infectious mononucleosis

Pathogenesis of bacterial meningitis in childhood

1. Invasion of CSF is preceeded by nasopharyngeal carriage of the organism
2. The bacteria which multiply in CSF are encapsulated and some share common surface antigens, e.g. *Esch. coli* K1, group b streptococcus type III and group B meningococcus have an antigenically-related surface polysaccharide
3. 75% of cases occur before the age of 5, during the period before development of antibodies. Males exceed females
4. Bacteraemia results in localisation in the subarachnoid space

5. Subsequent events include
   — formation of exudate
   — altered CSF permeablity
   — local antibody formation
6. Complications include
   — cerebral oedema
   — CSF obstruction
   — vasculitis, encephalitis
   — dural sinus thrombosis
   — cranial nerve palsies

Clinical features
1. Meningism
   Neck stiffness, photophobia, headache
2. Fever
3. Altered mental state
4. Focal signs and raised intracranial pressure
   — only if complicated or the presentation is late
5. Convulsions
   — are common in early childhood but usually due to fever rather than mengitis *per se*
6. Signs associated with specific infections:
   a. purpuric rash — meningococcus
   b. parotitis — mumps
   Diagnostic methods in central nervous system infections (see Table 75)

1. *CSF examination*
a. cell count and differential
b. Gram stain
c. culture — for bacteria and viruses
d. glucose, lactate levels
e. protein concentration
f. detection of antigens
   — using latex agglutination, coagglutination or immunoelectrophoresis
g. detection of endotoxin (Limulus lysate test)
h. special tests for certain agents
   Cryptococcosis — India ink stain
   Tuberculosis — acid fast stain
i. antibodies
   — can be detected in the CSF in more prolonged infections:
     (i) syphilis
     (ii) herpes simplex encephalitis
     (iii) coccidioidomycosis
     (iv) histoplasmosis
     (v) subacute sclerosing panencephalitis
     (vi) trypanosomiasis

2. Blood culture
— is positive in about 50% of cases of childhood bacterial meningitis

3. Peripheral white cell count and differential

4. Organ imaging techniques
— especially for brain abscess and encephalitis
— includes skull X-ray, computerised axial tomography

5. Electroencephalography
— is useful for encephalitis

6. Serology
— detection of changing or high antibody titres in peripheral blood
— especially useful in viral meningitis and encephalitis

'Aseptic' meningitis
This term, widely used for many years, implies acute meningitis with predominance of lymphocytes, normal glucose and no growth on bacterial culture. Important causes include:
1. Viral infections
2. Partially treated bacterial infections
3. Acute infections due to bacteria which do not grow on ordinary laboratory media (e.g. syphilis, leptospirosis, brucellosis)
4. Early stages of tuberculous and fungal meningitis
5. Non-infectious causes of meningitis (see below)

Causes of CSF pleocytosis other than intracranial infection
1. Malignancy
   — usually accompanied by presence of malignant cells
2. Cerebrovascular disease
3. Neurosurgical procedures
4. Hypoxia, including prolonged convulsions
5. Collagen-vascular disease
6. Infective endocarditis
   Staphylococcal septicaemia
   — may cause pleocytosis without invading CSF
7. Demyelinating diseases

Chronic meningitis
The syndrome is characterised by:
— signs of meningoencephalitis, often accompanied by increased intracranial pressure
— clinical progression over 3–4 weeks
— development of cranial nerve palsies
— CSF lymphocytosis, decreased glucose, increased protein

Causes
1. Infections
   a. tuberculosis
   b. fungi: especially cryptococcosis
   c. syphilis
   d. parasites
   e. brucellosis
   f. infective endocarditis
2. Non-infectious
   a. carcinoma, lymphoma
   b. systemic lupus erythematosis and other collagen-vascular diseases
   c. sarcoidosis

Special causes of meningitis
1. Contiguous infection from
   a. sinusitis
   b. compound fracture
   c. craniotomy
   d. mastoiditis
   e. ventricular drain
   f. CSF shunt
   g. dental abscess
   h. cavernous sinus thrombosis
   Common organisms causing contiguous infection include
   (i) *Streptococcus pneumoniae*
   (ii) *Staphylococcus aureus*
   (iii) *Staphylococcus epidermidis* (CSF shunts)
   (iv) Gram-negative rods (*Klebsiella, Enterobacter, Pseudomonas*)
2. Occupational and geographical associations with meningitis
   Amoebic meningitis
   — due to *Naegleria fowleri*, a free-living amoeba
   — follows swimming in freshwater pools
   — enters CSF through cribriform plate
   Eosinophilic meningitis
   — due to *Angiostrongyloides cantonensis*
   — occasionally due to other worms which enter CSF, viz,
      (i) *Taenia solium* (cysticercosis)
      (ii) *Gnathostoma spinigerum* (S.E. Asia)
   Tuberculous meningitis
   — in immigrants and travellers from countries with a high incidence of the disease
   Lymphocytic choriomeningitis
   — a virus associated with hampsters and mice
   Leptospirosis
   — associated with rats, domestic animals

Brucellosis
— associated with cattle, goats, pigs
Syphilis
— most common in male homosexuals
Lyme disease
— a recently defined tick-borne spirochaete (see chapter on Spirochaetes)
Streptococcus suis
— follows occupational exposure to pigs or pork
3. Meningitis associated with immune deficiency
   a. Listeriosis
   b. Cryptococcosis
   c. Chronic enteroviral infection
4. Neonates
   Special features of meningitis in this age group are:
   a. absence of the signs seen in older children and adults
   b. CSF cell count and glucose may be only slightly altered
   c. use of ampicillin and aminoglycosides rather than chloramphenicol for initial treatment
   d. pathogens derived from the various sources, as follows:

(i) Birth canal	*Streptococcus agalactiae* (group B streptococci) *Esch. coli*, capsular type K1 Herpes simplex, type 2 *Listeria monocytogenes*
(ii) Environment	Coliforms *Pseudomonas aeruginosa* *Flavobacterium meningosepticum*
(iii) Nursery epidemics	Enteroviruses, especially echovirus 11.
(iv) Transplacental	see Chapter on Infectious of the Female Genital Tract

5. Recurrent meningitis
   a. Anatomical defect
      — CSF leak
      — post-operative
      — neural tube defects
      — dermal sinuses
   b. Parameningeal focus
      — sinusitis
      — otitis media
      — epidural abscess
   c. Immune defect
      — complement deficiency
      — IgM deficiency
      — splenectomy
      — phagocyte defects

d. Non-infectious
   — epidermoid cysts
   — Mollaret's meningitis
   — collagen-vascular disease

Principles of management of meningitis

1. Antibiotic therapy should be intravenous (with the possible exception of chloramphenicol)
2. Duration of treatment should be adequate and for at least two weeks in most cases
3. Physical signs should be carefully charted, viz. temperature, mental state, pupils, fluid balance
4. Anticonvulsants
   — may be necessary for seizures, which complicate 10–40% of childhood meningitis and are mostly febrile fits rather than evidence of spread of infection
5. High fevers should be controlled
6. Raised intracranial pressure due to cerebral oedema may warrant treatment with fluid restriction, mannitol or steroids
7. Surgical procedures may be necessary for development of hydrocephalus, viz. decompression, ventricular drainage, CSF shunt
8. Subdural effusions may require repeated drainage

Antibiotics for meningitis

In the first years of life, meningitis suspected to be bacterial should be treated immediately with chloramphenicol.
Chloramphenicol
— Bactericidal for the three common causes of childhood meningitis
— Levels in CSF are about 50% of serum levels
— Given in a dose of 100 mg/kg/d (4–6 g/d in adults)
Ampicillin
— is about as effective as chloramphenicol for childhood meningitis but the emergence of resistant *H.influenzae* has eroded its value
— may replace chloramphenicol when sensitivity of the organism has been established. Dose is 200 mg/kg/d (12 g/d in adults)
Sulphadiazine
— was previously an agent of choice for *N.meningitidis*
Penicillin
— is the agent of choice for diagnosed pneumococcal and meningococcal infections
Aminoglycosides
— do not achieve adequate levels in CSF when given parenterally. Should be given intrathecally as well as systemically

Cephalosporins
— older agents (cephalothin, cephaloridine) are not useful for the treatment of meningitis
— new agents (cefotaxime, latamoxef) are effective for Gram-negative meningitis

Sequelae of meningitis
1. Subdural effusion or empyema
2. Venous sinus thrombosis
3. Cranial nerve palsies, deafness.
4. Mental retardation
5. Hydrocephalus

ENCEPHALITIS
Inflammatory disease of the brain substance due to:
1. Infection
2. Demyelination

Infectious causes include:
1. Viruses
   a. *Herpes simplex* is the commonest cause of encephalitis in Western countries
   b. *Arboviruses* (see chapter on RNA viruses)
   c. *Rabies* (see chapter on RNA viruses)
   d. *Enteroviruses*, especially poliovirus which causes a myelitis rather than encephalitis
   e. *Others*: encephalitis occasionally complicates common viral infections, due to tissue invasion or demyelination
2. Bacteria
   a. Listeriosis
   b. Syphilis
3. Protozoa
   a. Toxoplasmosis
   b. Trypanosomiasis
   c. Malaria
4. Metazoans
   a. Trichinosis
   b. Cysticercosis

Clinical manifestations of encephalitis
1. Fever
2. Meningism (if the infection involves meninges as well as the brain)
3. Alteration of consciousness
4. Focal neurological signs, seizures
5. Systemic findings, such as rash

Herpes simplex encephalitis
— occurs at all ages
— may be a primary or secondary infection
— causes a necrotizing encephalitis, predominantly involving the temporal lobes
— begins with fever, headache, bizarre behaviour, hallucinations
— followed by convulsions, hemiparesis, stupor
— CSF reveals pleocytosis, often with red cells
— EEG shows bilateral temporal lobe changes
— isotopic brain scan — reveals early change
   CT scan — reveals temporal lobe abnormality later than the isotopic scan
— antibodies appear in the CSF during convalescence
— treatment consists of adenine arabinoside or acyclovir

Poliomyelitis ("infantile paralysis")
— Poliovirus replicates preferentially in the neurones of the anterior horn cells
— Three antigenically distinct viruses are causative, making it possible for one subject to suffer three episodes of the disease
— Produce clinical disease as follows:
   1. Inapparent (subclinical) infection occurs in approximately 95% of cases. Virus is shed in stools, antibodies develop
   2. Prodromal illness — with fever, headache and malaise
   3. Aseptic meningitis — in about 1% of infections
   4. Poliomyelitis — follows 1–2 days after the prodromal illness in 0.1% of infections
      a. *Spinal*: Pain and tenderness in involved muscle groups followed by paralysis. Some recovery is the rule. Injections or exercise of muscles increases the risk of subsequent paralysis
      b. *Bulbar*: Most commonly involves cranial nerves 9 and 10, occasionally the respiratory and vasomotor centres. Tonsillectomy increases the risk of subsequent cranial nerve involvement
      c. *Cerebral* (polioencephalits). Rare, causes spasticity rather than flaccid paralysis

Diagnosis
1. Culture — readily yields the virus from throat or stool, rarely from CSF
2. Serology — may detect rising titres

Special causes of encephalitis
1. Opportunists
   a. *Listeria monocytogenes*
   b. *Toxoplasma gondii*

  c. Progressive multifocal encephalopathy
  d. Herpesviruses (CMV, HSV, VZ)
  e. Fungi (*Candida, Aspergillus*)
  f. *Strongyloides stercoralis*
2. Geographical, occupational
  a. Arboviruses
  b. Malaria
  c. Trypanosomiasis
  d. Rickettsiae
  e. Rabies
3. Slow viruses (see chapter on General Characteristics of Viruses)

BRAIN ABSCESS
Suppurative infection of brain tissue originates from:
1. Contiguous focus
  — otitis media, mastoiditis
  — paranasal sinuses
  — skull fracture, surgery
  Solitary lesions are the rule
2. Haematogenous
  — lung abscess, bronchiectasis
  — dental and oral infections
  — intra-abdominal infections
  — endocarditis
  Right-to-left cardiac shunts frequently predispose to blood-borne brain abscesses

Bacterial causes of brain abscess
1. *Oral flora*, particularly *Streptococcus milleri*, mixed aerobes and anaerobes
2. *Staphylococcus aureus*: despite the frequency of bacteraemia due to this organism it is an unusual cause of brain abscess
3. *Nocardia asteroides* follows pulmonary infection, often in immunosuppressed host

Clinical features of brain abscess
1. Headache and fever
2. Focal neurological signs, seizures
3. Raised intracranial pressure
4. Infection at contiguous or remote sites

Treatment of brain abscess
1. Drainage
2. Antibiotics (chloramphenicol, penicillin, metronidazole)

SOME INFECTIOUS CAUSES OF INTRACEREBRAL SPACE OCCUPYING LESIONS

1. Brain abscess
2. Listeria monocytogenes
3. Tuberculoma
4. Toruloma (cryptococcosis)
5. Cysticercosis
6. Trichinosis
7. Hydatids

Table 74 Investigation of intracranial infections

Note that lumbar puncture is dangerous in brain abscess and does not aid in the diagnosis

	Bacterial meningitis	Viral meningitis	Viral encephalitis	Brain abscess
CSF				
Cell count (per cu mm)	> 1000	< 500	5–50	Variable
Polymorphs	> 80%	< 50%	> 50%	> 50%
Protein (g/l)	> 100	50–100	50–100	Variable
Glucose (% of serum level)	< 50%	> 50%	>50%	> 50%
Lactate	Increased	Normal	Normal	Normal
Other CSF tests	Gram stain Culture	Virus culture	Antibodies (herpes encephalitis)	
	Bacterial antigen			
Blood culture	Positive in > 50%			
Serology	Not helpful	Epstein–Barr Mumps	Arboviruses	Not helpful
Peripheral white cell count	Increased	Normal or low	Variable	Increased
Brain scan	Normal	Normal	Late changes	Focal lesion with ring enhancement

Pulmonary infections

BRONCHITIS

Acute Bronchitis
Acute inflammation of the bronchial tree is due to:
1. Viruses causing common colds
2. Influenza
3. Adenoviruses
4. *Mycoplasma pneumoniae*
5. *Bordetella pertussis* (whooping cough)

Characterised by:
— cough and variable sputum production. Suprainfection by resident flora renders the sputum purulent
— low fever

Treatment is usually unnecessary

Chronic bronchitis
— is associated with continuous excessive secretion of mucus, due to cigarette smoking, occasionally atmospheric pollution
— may be complicated by airways obstruction and emphysema
— is prone to acute exacerbations due to
  1. Viruses and *Mycoplasma pneumoniae*
  2. *Streptococcus pneumoniae* and *Haemophilus influenzae*, which colonise the lower airways of these patients, periodically causing copious purulent sputum production and worsening airways obstruction. Bronchopneumonia may follow

Treatment: tetracyclines, ampicillin, cotrimoxazole

BRONCHIOLITIS

— Infection of the smaller airways due to
  1. Respiratory syncytial virus
  2. *Bordetella pertussis*
— Secondary infection may lead to bronchopneumonia

BRONCHIECTASIS

— Destructive dilatation of large bronchi following
  1. Severe pneumonia with suppuration, or
  2. Impaired pulmonary defence mechanisms, e.g. cystic fibrosis, immotile cilia syndromes
— Acute exacerbations have the same causes as chronic bronchitis
— Cystic fibrosis is also associated with *Staphylococcus aureus* and mucoid strains of *Pseudomonas aeruginosa*

Treatment:
1. Antibiotics for *H.influenzae* and *Strep.pneumoniae*
2. *Staph.aureus* may be treated with cloxacillin, fusidic acid, clindamycin
3. Physiotherapy (postural drainage)

PNEUMONIA

— Infection of the lung parenchyma beyond the terminal bronchioles
— Pneumonia has many causes, viz.

1. In the community
a. *Streptococcus pneumoniae*
b. *Mycoplasma pneumoniae*
c. Viruses
d. *Legionella pneumophila*
e. *Staphylococcus aureus*
f. *Mycobacterium tuberculosis*

2. In predisposed patients:
a. Chronic bronchitis (see above)
b. Aspiration — due to mixtures of aerobes and anaerobes
c. Postoperative, due to:
   (i) prior chronic bronchitis
   (ii) aspiration
d. Intensive care unit, following endotracheal intubation
e. Friedlander's pneumonia — in alcoholics
f. Bronchiectasis, cystic fibrosis

3. Pneumonia associated with occupational or geographical exposure to unusual agents (Table 76)

Table 75

Disease	Agent	Origin
Psittacosis	*Chlamydia psittaci*	Birds, especially psittacines
Q fever	*Coxiella burnetii*	Cattle
Melioidosis	*Pseudomonas pseudomallei*	S.E. Asia environmental
Plague	*Yersinia pestis*	Rodents
Anthrax	*Bacillus anthracis*	Cattle
Tularaemia	*Francisella tularensis*	Rodents
Cryptococcosis	*Cryptococcus neoformans*	Birds
Histoplasmosis	*Histoplasma capsulatum*	Soil and dust
Coccidioidomycosis	*Coccidioides immitis*	Soil of arid parts of California

4. Opportunistic (see chapter on Opportunistic Infections)

Clinical features of pneumonia
1. Fever, rigors, cold sores
2. Cough, tachypnoea, sputum production (occasionally haemoptysis)
3. Pleuritic chest pain, pleural friction rub
4. Signs of consolidation or effusion
5. Signs of underlying illnesses, e.g. alcoholism

Diagnosis
1. Culture of respiratory secretions
   a. *Sputum*: most easily obtained but always contaminated with oral flora
   b. *Pernasal swabs*: best diagnostic technique for young children
   c. *Gastric aspirate* is useful for:
      — neonates
      — after overnight fasting, for tuberculosis
   d. *Transtracheal aspirate*: avoids contamination by oral flora but risk of complications
   e. *Bronchial aspirate*: at bronchoscopy
      Endotracheal aspirate: following intubation
   f. *Lung biopsy or aspirate*, obtained percutaneously
      — best technique for obtaining uncontaminated specimens
      — risk of pneumothorax

2. Pleural aspirate or biopsy—if effusion is present
3. Blood culture
4. Serology
   — especially for viruses, psittacosis, Q fever, *Legionella*
5. Chest X-ray

Principles of management of pneumonia

1. Antibiotics
   — may be chosen on the basis of Gram stain examination of sputum but in community patients, the choice is usually based on clinical findings, viz.

   a. Rapid onset, high fever, pleurisy $\left\{\begin{array}{l}\text{Penicillin}\\\text{(Erythromycin)}\end{array}\right.$

   b. Underlying chronic bronchitis, copious purulent sputum $\left\{\begin{array}{l}\text{Tetracyclines}\\\text{Ampicillin, amoxycillin}\\\text{Cotrimoxazole}\end{array}\right.$

   c. Slow onset, persistent cough, familial $\left\{\begin{array}{l}\text{Tetracyclines}\\\text{Erythromycin}\end{array}\right.$

2. Physiotherapy
   Percussion, endotracheal aspiration, postural drainage
3. Humidified air
4. Oxygen—should be used carefully in the presence of airways obstruction
5. Rehydration
6. Bronchodilators—if infection complicates airways obstruction
7. Observations:
   Temperature, pulse, respiratory rate, blood gases, spirometry, sputum volume and nature
8. Follow-up chest X-ray is important to exclude malignancy and tuberculoisis

Features of specific causes of pneumonia

1. *Streptococcus pneumoniae*
   — sudden onset, sustained fever
   — prominent pleurisy and progression to consolidation
   — may be lobar or bronchopneumonic
   — most patients have underlying diseases, such as chronic bronchitis, alcoholism, influenza, diabetes, cardiac failure
   — small doses of penicillin suffice (1–2 g per day). Cephalothin, erythromycin and tetracycline are alternatives

2. *Mycoplasma* ('atypical') pneumonia
   — peak incidence at 5–15 years
   — occurs in families and enclosed populations
   — insidious onset of fever, cough, malaise
   — unremarkable physical signs, pleurisy is rare

— unilateral or bilateral patchy infiltrates in one or more
  bronchopulmonary segments
— treated with erythromycin or tetracycline. Relapse may occur if
  treated for less than 7 days
— unusual complications include haemolytic anaemia, meningitis,
  polyneuropathy

3. *Legionnaires' disease*
— more common in older males, smokers, immunosuppressed
— may be a history of exposure to excavations or air conditioners
— malaise and myalgia followed by high fever, pleurisy and signs
  of consolidation
— diarrhoea may occur at the onset
— bilateral involvement in 50%, lobes may be progressively
  consolidated
— treated with erythromycin, 0.5 g 6-hourly i.v. Rifampicin may be
  added.

4. *Staphylococcal pneumonia*
— involves the lung by one of two routes:
    a. *bacteraemic*: drug addicts, following intravenous infusions
    b. *descending*: following viral infections, especially influenza
— progression of symptoms is often rapid
— chest X-ray shows multiple opacities, which often become
  thin-walled cysts (pneumatocoeles), often followed by
  effusions, empyema, pneumothorax
— treated with flucloxacillin, methicillin, nafcillin or cephalothin,
  1–2 g i.v. every 4 hours
— empyemas may require drainage

5. *Viral pneumonia*
— usually preceeded by upper respiratory tract symptoms
— common causes include
    a. respiratory syncytial virus ⎫ in young children
    b. parainfluenzae, type 3   ⎭
    c. influenza
    d. adenoviruses — in military recruits
— typical chest X-ray reveals bilateral interstitial infiltrate
— is often followed by bacterial infection (*Strep. pneumoniae, H.
  influenzae, Staph. aureus*)

6. *Aspiration pneumonia*
— associated with:
    a. episodes of coma (epilepsy, stroke, etc)
    b. inhalation of foreign bodies
    c. bronchial malignancy
    d. oesophageal reflux
    e chronic bronchitis

— due to oral flora with anaerobes playing a leading role
— poor oral hygiene and periodontal disease often present
— lower and posterior lobes are usually involved (right more than left)
— onset is insidious
— cavitation and empyema are frequent
— treated with penicillin 6–12 Mu per day, often with metronidazole as well
— empyemas may require drainage

7. *Friedlander's pneumonia*
— a rare cause of pneumonia, almost always in alcoholics
— due to certain capsular serotypes of *Klebsiella aerogenes*
— more common in upper lobes, early cavitation, high mortality
— treated with cephalothin plus gentamicin

8. *Intensive care pneumonia*
— follows bacterial colonisation of the lower airways during prolonged endotracheal intubation
— Gram-negative rods, especially *Pseudomonas aeruginosa*, predominate
— preceded by tracheobronchitis. More common in chronic bronchitics
— fever, leucocytosis and segmental opacities suggest pneumonia but distinction from other intensive care causes of lung change (pulmonary oedema, adult respiratory distress syndrome, pulmonary infarction) may be difficult
— treatment according to sensitivity tests — usually cephalosporins, ticarcillin, aminoglycosides

Causes of pulmonary cavitation:
1. Infectious
   a. Aspiration
   b. Tuberculosis
   c. *Staphylococcus aureus*
   d. Friedlander's pneumonia
   e. Fungi, nocardia, actinomycosis
   f. Pneumococcus — is a rare cause of cavitation
2. Neoplasm
3. Vasculitis
   Wegener's granulomatosis, polyarteritis nodosa
4. Pulmonary infarction

Causes of unresolved pneumonia
1. Aspiration (especially with foreign body), including lipoid pneumonia
2. Malignancy, bronchial obstruction

3. Tuberculosis
4. Fungus
5. Actinomycosis

Causes of recurrent pneumonia
1. Anatomical defects (the pneumonia recurs in the same lobe)
   a. Bronchiectasis
   b. Bronchial obstruction especially neoplasm
   c. Sequestrated segment
2. Immune deficiency

Enteric infections

Definitions of clinical syndromes
1. *Food poisoning*: due to ingestion of preformed toxin. Incubation period less than 12 hours and symptoms rarely last longer than 24 hours
2. *Gastroenteritis, enteritis, enterocolitis*: multiplication of pathogen in small and often large bowel. 'Gastroenteritis' is a misnomer — despite vomiting, the stomach is not involved. Incubation period is 1–3 days
3. *Dysentery*: infection with predominant colonic symptoms (colitis) or rectal symptoms (proctitis)

FOOD POISONING
1. Abrupt onset, clear relationship to food, often epidemic or institutional.
2. Causes
   a. *Staphylococcus aureus*: custards, puddings, potato salads. Vomiting is predominant (due to action of toxin on nervous system).
   b. *Clostridium welchii*, type A: poorly-cooked meat or gravy. Diarrhoea and colicky abdominal pain 6–12 hours after the meal.
   c. *Bacillus cereus*: frequently associated with rice. Produces two toxins, one causes vomiting, the other diarrhoea.
   d. *Clostridium welchii*, type C: follows ingestion of poorly-cooked pork, in highlands of New Guinea ('pig-bel'). Causes small bowel necrosis. High mortality
   e. *Clostridium botulinum*: Botulism
   f. Scombrotoxin, shellfish toxins, ciguatoxin, mushroom toxins: cause neurological symptoms and altered mental state
   g. Heavy metals, bromates, detergents, hydrocarbons: cause vomiting

ENTERITIS

Pathogenesis — 3 mechanisms:
1. Toxigenic. following mucosal attachment of bacteria, with release of toxin. No mucosal damage (e.g. *Vibrio cholerae*)
2. Invasive. following bacterial invasion of the mucosa or submucosa (e.g. *Shigella*, Salmonella). (Some bacteria may act via both of the above mechanisms)
3. Damage to small bowel enterocytes: causing brush border enzyme depletion (? viruses, ? *Giardia lamblia*)

Factors influencing gut colonisation by pathogens
1. Inoculum size
2. Gastric acidity
3. Peristalsis
4. Normal bowel flora
5. Immunity
   a. Active, from prior exposure
   b. Passive, from breast feeding

Enterotoxigenic enteritis

1. Clinical features
a. Incubation period about 24 hours
b. Duration of symptoms 1–2 days
c. Fever low or absent
d. Colicky abdominal pain, watery diarrhoea
e. No mucosal damage, no faecal leucocytes

2. Causative organisms
a. *Vibrio cholerae*
b. *Escherichia coli* (travellers' diarrhoea)
c. Other Gram-negative bacteria (*Aeromonas*, *Vibrio* species, ? *Klebsiella*)

3. Pathogenesis
a. Attachment of bacteria to small bowel mucosa followed by multiplication and toxin production
b. Toxin attaches to cell wall gangliosides stimulating adenylcyclase activity. Cellular cylic AMP level rises
c. Increased transport of C1, followed by Na and water, into the bowel lumen

4. Characteristics of V. cholerae exotoxin
a. protein of molecular weight 83,000, consisting of two subunits
b. pH optimum, 8
c. antigenic, heat labile
d. binds to GM1 ganglioside of mucosal cell

Enterotoxigenic Esch. coli has two toxins (see Table 77)
a. LT (heat labile): very similar to V. cholerae toxin
b. ST (heat stable): smaller, less antigenic, stimulates cyclic GMP
    production

Methods for determing enterotoxigenicity of Esch. coli
1. Using culture filtrate
    a. ligated rabbit ileal segment
    b. tissue cultures (adrenal tissue, Chinese hamster ovary)
    c. immunoassay
2. Detection of plasmid bearing enterotoxin gene using DNA probes

Invasive enterocolitis

1. Clinical features:
a. Incubation period 1–3 days
b. Duration 5–7 days
c. Fever is usually present
d. Colicky lower abdominal pain
e. Diarrhoea is less copious, occasionally contains blood and
    mucus. Faeces contain numerous leucocytes

2. Causative organisms
a. *Salmonella* species
b. *Shigella* species
c. *Campylobacter fetus*, subspecies *jejuni*
d. *Yersinia enterocolitica*
e. *Escherichia coli* (invasive strains)
f. *Entamoeba histolytica*

3. Pathogenesis
a. *Salmonella*: invades submucosa. Probably also produces
    enterotoxin.
b. *Shigella*: invades and destroys mucosal cells.

Viral enteritis

1. Rotavirus:
a. Peak incidence at 6–24 months of age ('weanling diarrhoea')
b. More common in cooler months of the year
c. Incubation period 1–2 days
d. Fever, vomiting and diarrhoea lasting 4–5 days, often causing
    dehydration
e. Cough, nasal discharge and otitis media are often associated
f. Rotavirus infections in neonates and adults are usually
    asymptomatic

Table 76 Three types of *Esch. coli* causing enterocolitis

	Enterotoxigenic Esch. coli (ETEC)	*Enteropathogenic Esch. coli (EPEC)*	*Enteroinvasive Esch. coli (EIEC)*
Antigenic type	Any serotype	Certain O serotypes esp. 28, 55, 111, 112	O157:H7 (recently reported)
Pathogenesis	1. Produce two enterotoxins: LT: heat-labile high M.Wt. activates adenylcyclase ST: heat-stable low M.Wt. activates guanylcyclase 2. Adhesion (colonisation) factors	Colonisation of small bowel Adhere to Hep2 cells The mechanism by which EPEC cause diarrhoea is poorly understood	Penetration and multiplication with in intestinal epithelial Positive Sereny test Cytotoxic Identical to toxin of *Shigella dysenteriae* I
Significance	1. Diarrhoea in travellers 2. Infantile diarrhoea	Epidemics in neonatal nurseries	Food-borne epidemics have been recently described
Clinical features	Watery diarrhoea	Watery diarrhoea	Haemorrhagic colitis

2. Parvovirus-like agents (Norwalk, Hawaii, Cockle, Parramatta agents, etc)
a. Occurs most frequently in winter months, all ages are affected
b. Epidemics may be associated with food, water or shellfish
c. Incubation period is 24–48 hours, disease lasts 2–3 days
d. Vomiting, diarrhoea, low fever

3. Others
Other viruses found in faeces have uncertain roles as causes of diarrhoea (echoviruses, coxsackieviruses, adenoviruses, caliciviruses)

COLITIS

Causative organisms
1. *Shigella*: multiply initially in small bowel causing pain and fever, later invade and damage colonic mucosa causing dysentery.
2. *Salmonella, Campylobacter, Yersinia*: all involve colon to a variable degree, as well as small bowel
3. *Entamoeba histolytica* (amoebiasis): invades large bowel at sites of stasis (caecum and sigmoid). Insidious onset, prolonged colonic symptoms, 2–3 bowel actions per day. Often blood and mucus.
4. *Clostridium difficile*: causes pseudomembranous and antibiotic-associated colitis by toxin formation. Almost always follows antibiotic therapy. Bowel stasis exacerbates
5. *Schistosoma mansoni* and *japonicum*

PROCTITIS

As well as being caused by the agents of colitis, infection involving the rectum in male homosexuals may be sexually transmitted, as follows:
1. Herpes simplex
2. Syphilis
3. Gonorrhoea
4. Lymphogranuloma venereum
5. Donovanosis
6. Campylobacter-like organisms

DIAGNOSTIC METHODS IN INFECTIOUS ENTEROCOLITIS

1. Stool examination
Numerous techniques may be required to detect all causes:
a. Microscopic examination for:
   — leukocytes (indicate invasive disease)
   — eggs, cysts, trophozoites, larvae (following stool concentration)

— modified acid-fast stain for *Cryptosporidia*
— darting motility of *Campylobacter fetus*

b. Culture, using selective media for
— *Salmonella* and *Shigella*
— *Campylobacter*
— Vibrios
— *Clostridium difficile*
— *Yersinia*

c. Toxin detection
— *C. difficile* (cytotoxic for tissue cultures)
— Enterotoxins (see above)

d. Antigen detection
— Rotavirus

2. Sigmoidoscopy and rectal biopsy
— is useful for invasive parasitic disease of colon (amoebiasis, schistosomiasis)
— is an important diagnostic technique for non-infective colonic disease especially ulcerative colitis and Crohn's disease

3. Duodenal aspirate and biopsy
— detects small bowel parasites (*Giardia, Strongyloides, Cryptosporidium*)
— diagnostic changes in Whipple's disease

4. Barium enema
— non-diagnostic changes in amoebiasis, pseudomembranous colitis

5. Serology
— useful for amoebiasis, schistosomiasis, strongyloidiaisis and yersiniosis

Gut infections causing steatorrhoea:
1. *By colonising brush border*: Giardia, Strongyloides, Cryptosporidium
2. *By damaging lamina propria*: Whipple's disease (the causative agent has not been identified)
3. *By causing overgrowth of bacteria in small bowel*: Blind loops, strictures, strictures, stasis, hypomotility. Tropical sprue

PRINCIPLES OF MANAGEMENT OF ENTEROCOLITIS

1. Fluid and electrolytes
a. Assess and replace fluid volume by intravenous and/or oral routes
b. Maintain fluid balance charts
c. Check serum electrolytes and acid-base balance. Correct accordingly

2. Antidiarrhoeal agents
— may prolong disease. Use sparingly in adults and not at all in young children
  a. opiates: codeine, diphenoxylate, loperamide
  b. kaolin, aluminium hydroxide

3. Antimicrobial agents
— are only useful for some infections
  a. *Salmonellosis*: not indicated. May prolong the symptoms and the excretion of the organism
  b. *Shigellosis*: ampicillin, trimethoprim compound, nalidixic acid
  c. *Cholera*: tetracyclines
  d. *Amoebiasis* ⎰
     Giardiasis ⎱ metronidazole, tinidazole
  e. *C. difficile*: vancomycin, bacitracin, metronidazole
  f. *Campylobacter*: erythromycin
  g. *Esch. coli*: neomycin, streptomycin

4. Prevention of spread:
  a. Appropriate disposal of faeces
  b. Handwashing
  c. Exclusion from food handling

PARASITIC INFESTATIONS CAUSING DIARRHOEA:

1. Protozoans
   — Amoebiasis
   — Giardiasis
   — Cryptosporidiosis
2. Metazoans
   — Strongyloidiasis
   — Schistosomiasis
Note that most parasites inhabiting the gut lumen do not cause diarrhoea.

Amoebiasis
— results from invasion of the gut by *Entamoeba histolytica* at sites of bowel stasis (caecum and sigmoid), resulting in ulceration and stricture formation
— usually has an insidious onset with 2–3 bowel actions per day, accompanied by blood and mucus
— is transmitted by cysts in food and water, occasionally by contact
— is diagnosed by detection of red cell- containing trophozoites in freshly collected unformed stools or by detecting high titres of circulating antibody
— responds to metronidazole 400 mg q.i.d. p.o.

Giardiasis
— results from colonisation and attachment of trophozoites of
   Giardia lamblia to the mucosa of the duodenum and upper
   jejeunum. Enterocytes are damaged
— is transmitted by contact or by drinking water
— causes sudden onset of watery diarrhoea, upper abdominal pain
   and distension, flatulence
— later becomes subacute with loose stools, sometimes
   malabsorption and weight loss
— severe and prolonged disease may be associated with IgA
   deficiency
— is diagnosed by finding trophozoites in unformed stools or cysts
   in formed stools (repeated examination may be necessary)
— in some patients, duodenal aspiration may be necessary
— is treated with metronidazole 200–400 mg t.d.s. p.o.

Hepatitis

Most infections of the liver are viral (see Table 78)
1. Hepatitis A
2. Hepatitis B
3. Hepatitis B complicated by delta agent
4. Hepatitis, non-A, non-B
5. Infectious mononucleosis
6. Cytomegalovirus
   Rare causes of hepatitis include:
1. Herpes simplex (in the immunosuppressed)
2. Viruses causing haemorrhagic fevers (yellow fever, dengue, lassa virus)
3. Q fever
4. Leptospirosis

HEPATITIS A VIRUS (HAV)

— Recently designated enterovirus 72, with morphology and epidemiology typical of the group
— Virion was identified in 1973, as 27 nm particles in stools of patients. Recently grown in tissue culture
— Causes hepatitis in some primates as well as man
— In the USA about 20% have antibodies by adulthood, compared to 90% in developing countries
— Incubation period 14–50 days (average 30)
— Persistent infection and chronic liver disease do not occur
— Aplastic anaemia may rarely follow hepatitis A infection

Diagnosis
1. Detection of circulating IgM antibodies in the month following the onset of symptoms (using RIA, ELISA)
2. HAV particles can be detected in faeces for 2–3 weeks before the onset of jaundice and for about a week afterwards, but not a practical method for making the diagnosis

Table 77 A comparison of three causes of viral hepatitis

	Hepatitis A	Hepatitis B	Hepatitis nonA, non B
Nucleic acid	RNA	DNA	Unknown
Particle size	27 nm	42 nm	Unknown
Family	Picornavirus	Hepadnavirus	Unknown
Transmission	Faeco-oral	Ingested secretions Parenteral	? Ingested Parenteral
Incubation period (days)	14–50 av. 30	40–160 av. 90	15–160 av. 50
Population at risk	Children Travellers	Hospital workers Blood recipients Drug addicts Male homosexuals Infants of carrier mothers	Blood recipients Drug addicts
Carrier state	Nil	0.1–1.0%	Occurs but incidence unknown
Sequelae	Nil	Chronic active hepatitis Cirrhosis Hepatoma	Cirrhosis
Passive immunisation	Pooled adult gammaglobulin	Hyperimmune gammaglobulin	Not available
Active immunisation	Not available	Surface antigen	Not available

HEPATITIS B VIRUS (HBV)

— One of four viruses now in a recently created class called hepadnaviruses.
— The virus was first identified in serum following the discovery of three particle types
  1. Small round particles 15–25 nm diameter
  2. Tubules of 20 nm diameter and length 50–200 nm
  3. Larger spherical particles (Dane particles) of 42 nm diameter (28 nm core, inner shell 2 nm and outer coat of 7 nm)
— Unlike hepatis A virus, HBV causes more severe hepatitis and progressive disease.

Complications of hepatitis due to HBV
1. Persistent HBs antigenaemia
2. Fulminant hepatitis
3. Chronic persistent hepatitis
4. Chronic active hepatitis
5. Cirrhosis
6. Hepatoma
— Diagnosis is made by the detection of circulating antigens and antibodies ('markers') (Table 79)

Table 78

HBs Ag	Present before and after the onset of clinical disease in about 90% of cases and may persist for months or years.
anti-HBs Ag	Appears a variable period of time after HBs Ag disappears. Persists lifelong. Immune complexes with HBs Ag occuring during incubation period may cause arthritis, vasculitis
HBc Ag	Cannot be detected free in the circulation.
anti-HBc Ag	Appears early in the course of infection and persists lifelong. May be the only marker detectable in both acute and chronic disease, in which case IgM antibodies are detectable.
HBe Ag	Present in acute hepatitis. Prolonged detection is associated with infectivity and chronic liver disease.
anti HBe Ag	Appears 2–6 weeks after disappearance of HBe Ag, signalling recovery from either acute or chronic infection.

Risk factors for acquiring infection with HBV
1. Contact with blood or Hospital staff
   its products Recipients of multiple blood transfusions
   Drug addicts (via shared needles)
2. Sexual transmission Sexual contacts of carriers
   Male homosexuals
3. Contact with oral Dentists
   secretions Inmates of institutions

DELTA AGENT

— A virus-like particle consisting of the delta antigen and RNA enclosed in a particle coated with hepatitis B surface antigen
— Depends upon hepatitis B for replication (i.e. it is a defective virus)
— Transmission is similar to that of hepatitis with outbreaks reported amongst drug addicts and haemophiliacs
— Endemic in Southern Italy and the Mediterranean. Uncommon in Asia

— Superinfection of hepatitis B carriers causes
  1. Fulminant hepatitis
  2. Subacute hepatitis progressing rapidly to cirrhosis over
     several years
— Diagnosed by presence of IgM anti-delta antibodies

HEPATITIS, NON A, NON B

— A term used for hepatitis due to undefined viruses
— As judged by epidemiological studies, at least two transmissible
  agents may be involved. Transmission to chimpanzees has been
  achieved
— About 25% of sporadic hepatitis is not due to HAV, HBV, CMV or
  EBV
— The range of incubation period is unusually wide (35–70 days)
  and tends to fall between HAV and HBV
— Clinical disease is usually mild but chronic active hepatitis and
  cirrhosis are frequent if infection follows blood transfusion and
  abnormal liver function tests persist for more than one year in
  10–60%
— 6–35% of blood transfusions, depending on the donor source,
  are followed by sustained elevations of plasma transaminases
— Patterns of transmission appear to be similar to those of HBV

CLINICAL FEATURES OF HEPATITIS

1. Anorexia, nausea and vomiting, distaste for cigarettes
2. Jaundice, dark urine, pale stools, prurutis
3. Right upper quadrant abdominal pain and tenderness

BIOCHEMICAL CHARACTERISTICS OF HEPATITIS

1. Serum transaminases (aminotranferases) usually exceed 800 IU
   at the height of the illness
2. Serum bilirubin may continue to rise after transaminases are
   falling. Prolonged high levels indicate severe disease. Total
   bilirubin is equally divided between conjugated and
   unconjugated fractions
3. Serum alkaline phosphatase is normal or mildly elevated

GENERAL PRINCIPLES OF MANAGEMENT OF HEPATITIS

1. Bed rest is advocated during the symptomatic period but does
   not hasten the resolution of histological changes
2. Diet does not influence the outcome but low fat, high protein food
   is usually advocated
3. Hepatotoxic drugs, including alcohol, should be avoided

4. Isolation from other patients and care with the disposal of secretions and excreta is required (see chapter on Hospital Infections)
5. Immunisation of contacts
6. Treatment of liver failure
7. Monitoring of biochemical parameters and antigenaemia
8. Liver biopsy — is indicated if there is doubt about the diagnosis or if HBs Ag is still present and transaminases remain elevated 6–12 months after onset of infection
9. Steroids may be used for prolonged cholestasis but are contraindicated in chronic active hepatitis

Infections of bones and joints

OSTEOMYELITIS

Two routes exist for infection of bone:
1. Haematogenous (via the circulation)
2. Contiguous (from an adjacent site of infection)

Pathogenesis of haematogenous infection
1. Most often involves the rapidly growing long bones of prepubertal children, more commonly in males. Local trauma may precede.
2. The portal of entry is either unknown or a furuncle
3. Bacteria lodge in capillary loops
   a. *in children*, mostly the metaphyses of long bones (especially femur, tibia, humerus)
   b. *in adults*, mostly the bone plates of vertebral bodies adjacent to the intervertebral disc
4. Infection spreads laterally, perforating the cortex, lifting the loose periosteum and creating a subperiosteal collection. Between the ages of 1 year and puberty, the epiphysis is protected from infection
5. Growth of new bone under the elevated periosteum creates the involucrum
6. Infection in bone extends through the Haversian and Volkmann canals, vascular occlusion follows and large segments of bone necrosis occur (sequestrum formation)
7. Spontaneous drainage of infection on to the skin surface may result in permanent or recurrent sinus formation
8. In adults, infection of long bones may result in circumscribed bone destruction with a cavity surrounded by sclerotic bone (Brodie's abscess)

Sites of haematogenous osteomyelitis
1. Long bones (femur, tibia, humerus)
2. Vertebrae (in adults)

Organisms responsible for haematogenous osteomyelitis
1. *Staphylococcus aureus* by far the commonest cause
2. *Streptococci* Strep. pyogenes, *Strep. pneumoniae*
3. *Coliforms* mostly in adults, secondary to urinary tract infection

Clinical features
1. Fever, rigors
2. Local pain, tenderness, swelling and limitation of movement
3. In adults with vertebral osteomyelitis, the onset may be insidious

Diagnostic methods
1. Blood culture is positive in 50–60% of acute osteomyelitis in childhood
2. High white cell count and high ESR is the rule
3. Isotopic scans (technetium, gallium) reveal increased uptake within 2–3 days of the onset
4. Radiological change (destruction, cortical irregularity, sequestrum formation, periosteal reaction) do not appear in less than 10 days after the onset
5. Needle aspiration of subperiosteal pus or bone

Principles of management
1. Every effort should be made to establish a bacteriological diagnosis (blood culture, aspiration, operative specimens)
2. Appropriate antibiotics (usually antistaphylococcal) should be given in adequate dose for a least 1 month
3. Surgery is indicated if there is evidence of subperiosteal abscess formation

ARTHRITIS

Joint infection has many causes

1. Acute bacterial
a. *Staphylococcus aureus*: the commonest cause after early childhood
b. *Haemophilus influenzae*: the commonest cause in the first 2 years of life
c. *Neisseria gonorrhoeae*: the commonest cause in young adults
d. Streptococci: *Strep. pyogenes, Strep. pneumoniae*
e. Gram-negative rods: usually follows urinary tract infection

2. Subacute bacterial
a. Tuberculosis
b. Lyme disease
c. Syphilis

3. Viral
a. *Rubella*, especially in young women
b. *Other togaviruses* (Ross River, O'nyong-nyong, Chikungunya viruses)
c. *Hepatis B virus*, due to immune complex deposition at the onset of disease

4. Fungal
   a. *Systemic mycoses*: (coccidioidomycosis, histoplasmosis, sporotrichosis)
   b. *Candida*: following venous access

Factors predisposing to septic arthritis
1. Pre-existing joint disease, especially rheumatoid arthritis
2. Entry of organisms at sites of venous access
   — intravenous heroin users (*Staph. aureus*, *P.aeruginosa*, *Candida*)
   — prolonged use of venous lines
3. Steroid administration
4. Intra-articular injection

Clinical features
1. Pain and limitation of joint movement
2. Synovial thickening, effusion
3. Fever

Diagnosis
1. Aspiration and examination of synovial fluid, as follows:
   a. appearance
   b. cell count and differential. In septic arthritis, cell count usually exceeds 100,000 µl with more than 90% polymorphs
   c. Gram stain
   d. Culture
   e. Lactate/glucose
   f. Microscopy with polarised light (for crystals)
2. Culture of other relevant sites
   a. Blood
   b. Urine
   c. for *N.gonorrhoeae*: cervix, urethra, rectum, pharynx
3. Serology
   — for rubella, Ross River virus, HBs Ag, rheumatoid factor, antinuclear antibodies
4. Arthroscopy, synovial biopsy
   — may be warranted for suspected tuberculous arthritis
The differential diagnosis of inflammatory joint disease includes:
1. Rheumatoid arthritis, other connective tissue diseases
2. Crystal arthropathy (uric acid, calcium pyrophosphate)
3. Reactive arthritis
4. Degenerative disease

Principles of management of septic arthritis
1. Antibiotics
2. Drainage
3. Immobilisation

Gonococcal arthritis
— occurs mostly in females
— usually accompanies menstruation; the primary infection is often asymptomatic
— is associated with two clinical presentations
   a. fever, polyarthritis, tenosynovitis, vasculitic skin lesions. Blood culture may be positive
   b. monoarticular arthritis with organism present in synovial fluid

Infection in joint prostheses
Orthopaedic prostheses implanted with the use of acrylic cement are at risk of a slowly developing infection which has its origin in:
a. operative contamination with skin organisms, or
b. haematogenous infection
   Criteria for making the diagnosis include:
1. Pain and tenderness
2. Draining wound or sinus communicating with joint
3. ESR > 30, leucocytosis
4. Radiological changes: loosening, bony resorption, arthrogram revealing contrast agent between cement and bone
5. Isotopic scan revealing increased uptake
6. Joint aspirate revealing bacteria
7. Purulent joint fluid and synovial inflammation at surgery

REACTIVE ARTHRITIS
— Acute non-purulent arthritis following urethritis or enteric infections in genetically susceptible individuals
— Includes Reiter's syndrome (a triad of non-specific urethritis, conjunctivitis and arthritis) as well as similar syndromes
— In most patients, tissue typing reveals the presence of the HLA B27 antigen
— Associated infections include
   1. Sexually acquired urethritis
      a. *Chlamydia trachomatis*
      b. *Ureaplasma urealyticum*
   2. Enteric infections
      a. *Shigella* species
      b. *Salmonella* species
      c. *Yersinia* species
      d. *Campylobacter* species
      e. *Clostridium difficile*

Ear, nose, throat and eye infections

UPPER RESPIRATORY TRACT INFECTIONS

1. Common cold (Coryza)
— Viral infection of the upper respiratory tract characterised by:
   a. catarrh (sneezing, nasal discharge and obstruction)
   b. sore throat
— Several viruses are causative, especially rhinoviruses, but the aetiology of at least 40% of cases cannot be established
— Bacteria may later invade sites of mucosal damage including sinuses, middle ear and trachea
— Symptomatic treatment includes vasoconstrictors (decongestants), cough suppressants, analgesics, antihistamines

2. Pharyngitis
— Infection of the pharynx producing discomfit and inflammatory change
— Has bacterial and viral causes, viz.
   a. *Viruses causing colds*: by far the commonest cause
   b. *Adenoviruses* cause more severe symptoms plus fever. Conjunctivitis may be associated
   c. *Herpangina* due to Coxsackieviruses. Vesicles followed by small ulcers on posterior pharynx. Fever
   d. *Infectious mononucleosis* may be difficult to distinguish from other causes of pharyngitis
   e. *Streptococcus pyogenes* is the most common bacterial cause of pharyngitis, characterised by fever, exudate covering tonsils and posterior pharynx, tender lympadenopathy
   f. *Peritonsillar abscess* (quinsy) due to *Strep. pyogenes*, *Staph. aureus* or mixed anaerobes
   g. *Vincent's angina*, due to oral spirochaetes and Fusobacteria
   h. *Diphtheria*
   i. *Mycoplasma pneumoniae*
   j. *N. gonorrhoeae* acquired venereally. Rarely symptomatic

Investigation of upper respiratory tract infections

a. *Gram stain* is rarely useful for the diagnosis of infections in the oral cavity with the exception of:

Vincent's angina   — reveals a mixture of spirochaetes and fusiform bacteria

Candidiasis   — reveals pseudomycelium in scrapings obtained from typical plaques

b. *Bacterial culture* especially for:

Strep. pyogenes: Blood agar. Anaerobic culture enhances haemolysis.
Sheep blood prevents growth of haemophili.

C. diphtheriae: Tellurite agar and Loeffler's serum slope.

N.gonorrhoeae: Thayer-Martin medium, or similar.

Candida albicans: This yeast can usually be recognised on blood agar. Sabaroud's medium is selective.

c. Serology is useful for the following:

Strep. pyogenes: Antistreptolysin O, anti streptococcal DNA-se

Epstein-Barr virus: Heterophile antibodies (Paul-Bunnell test)

Mycoplasma   ⎰ CFT's are available but not usually employed for
Adenoviruses   ⎱ pharyngitis due to these agents.

d. *Virus culture*, using swabs or washings. Expensive

e. *Immunofluorescent antibody tests* using specific antiserum can be made on smears obtained via pernasal swabs

3. Tonsillitis

Term used for pharyngitis predominantly involving the tonsils and faucial pillars. Commonest causes are:

a. respiratory viruses
b. *Strep. pyogenes*
c. infectious mononucleosis

4. Rhinitis

Inflammatory change in nasal mucosa. Causes include:

a. common cold (see above)
b. allergy
c. atrophic — associated with *Klebsiella ozaenae*, although this is probably an opportunistic invader
d. granulomatous (scleroma) — caused by *Klebsiella rhinoscleromatis*, occurs in foci in Europe, Africa and Asia

5. Stomatitis
Ulcerative or necrotising involvement of the oral cavity rather than the pharynx. Has many causes, both infectious and non-infectious:
a. Viruses
— herpes simplex
— herpangina (Coxsackie A viruses)
— viral enanthems: measles, varicella, smallpox
b. Aphthous stomatitis
c. Neutropenia
d. Candidiasis
e. Drugs
f. Vitamin deficiency

6. Glossitis
Inflammatory diseases of the tongue. Has the same causes as stomatitis

7. Sinusitis
— Inflammation of the paranasal sinuses (maxillary, frontal, ethmoid and sphenoidal)
— Mucosal swelling accompanied by increased mucus secretion
— More common after adolescence when sinuses are fully developed

Pathogenesis
— begins as a viral infection in most cases
— exacerbated by ostial obstruction due to allergy, septal deviation, foreign bodies and polyps
— occasionally follows dental infection penetrating the floor of maxillary sinus

Microbiology of sinusitis
Respiratory viruses usually initiate acute sinusitis
Strep. pneumoniae ⎰ are the bacteria which most frequently
H.influenzae ⎱ superinfect viral sinusitis
Mixed anaerobes occasionally predominate in chronic disease
Staph. aureus is uncommon but occasionally associated with osteomyelitis (e.g. Pott's puffy tumour)
Coliforms ⎰may be present in chronic sinusitis, especially when
P. aneruginosa ⎱ associated with bronchiectasis
Aspergillus is occasionally associated with chronic disease
Phycomycetes cause mucormycosis, a rare complication of uncontrolled diabetes and prolonged neutropenia

Clinical features, diagnosis and therapy
a. Nasal obstruction and discharge, voice change
b. Facial pain, especially in the morning. Facial tenderness
c. Transillumination of maxillary sinus may reveal dullness
d. X-rays reveal opacity or fluid level
e. Lavage or sinoscopy may be useful diagnostically
f. Culture of nasal secretions is non-contributory
g. Complications (unusual) include osteomyelitis, brain abscess, meningitis, cavernous sinus thrombosis
h. Treatment includes
   — antibiotics (usually penicillins, trimethoprim compound, tetracyclines, erythromycin)
   — decongestants, steam inhalation
   — surgery for deviated septum, polyps. Occasionally maxillary antrostomy (Caldwell-Luc procedure)

8. Laryngitis, laryngotracheobronchitis
Infection of the larynx and lower respiratory tract causing cough and hoarseness. Almost always viral.

9. Croup (see Table 79)
— Acute laryngotracheobronchitis following viral upper respiratory tract infection in young children
— Inflammation at the narrow subglottic level produces characteristic stridor, barking cough, hoarseness and variable degrees of fever
— May be complicated by hypoxia, tachypnoea, hypercapnoea
— Course may fluctuate. Observation is the rule

10. Epiglottitis
— Rapidly progressive cellulitis of the epiglottis and adjacent structures nearly always due to *H.influenzae*, type b
— Mostly in children, age 2–4, occasionally in adults
— Causes fever, stridor, respiratory distress, dysphagia, drooling
— Tongue depressor reveals cherry-red epiglottis. X-ray may confirm
— Blood cultures are usually positive
— Tracheostomy is required, together with intravenous ampicillin or chloramphenicol. See Table 79.

OTITIS EXTERNA
— Infection of the squamous epithelium or skin appendages of the external auditory canal
— Normally resists infection because of lateral migration of keratin and debris, wax formation, low pH

Table 79 Comparison of three infections of the upper respiratory tract leading to airways obstruction in children.

	Epiglottitis	Croup	Laryngeal diphtheria
Incidence	Uncommon	Common	Rare
Cause	H.influenzae	Respiratory viruses, esp. parainfluenza	C.diphtheriae
Age	2–4 years	3 months–3 years	5–14 years
Site	Epiglottis	Subglottis (encircled by cricoid cartilage)	Extension of pharyngeal disease to the larynx
Prelude	Usually nil	Rhinitis, pharyngitis	Sore throat
Season	Summer	Winter	None
Onset	Rapid	Slow	Variable
Clinical features	Stridor Dysphagia Drooling High fever	Stridor Barking cough Hoarseness Variable fever Tachypnoea	Stridor Sore throat Toxaemia
Examination	Cherry red epiglottis	Pharyngitis	Membrane on posterior pharynx
Diagnosis	Blood culture	Virus culture	Throat swab and culture
Treatment	Ampicillin Chloramphenicol Endotracheal intubation in all cases with stridor	Humidification Close observation Blood gases Oxygen	Antitoxin Erythromycin or penicillin Tracheostomy Bronchoscopy to remove membrane
Progress	Rapid and unpredictable airways obstruction. Mortality rate is 6–25% without intubation	Fluctuating with eventual spontaneous recovery	Variable
Prevention	Nil	Nil	Vaccine (toxoid)

— Causes of infection include:
 1. Chronically wet external ear canal
    — due to high humidity, swimming, repeated trauma,
      seborrhoea, eczema, exostoses, allergy, and otitis media
      with drum perforation
    — encourages the growth of gram negative bacteria,
      especially *Pseudomonas aeruginosa*
    — causes pruritis, discharge, mild pain
    — treated by cleaning, drying, topical antibiotics
    — unusual variants include
      a fungal otitis externa due to *Aspergillus niger*
      b. malignant otitis externa in elderly diabetics causing
         osteomyelitis, severe pain and cranial nerve palsies
 2. Furunculosis due to *Staph. aureus*
    — causes severe pain and swelling, eventually discharge
 3. Bullous myringitis (infection of the tympanic membrane)
    — is due to viruses and *Mycoplasma pneumoniae*
    — causes rapidly developing pain, blebs on the tympanic
      membrane, haemoserous discharge
 4. Herpes zoster (Ramsay–Hunt syndrome)
    — may involve geniculate ganglion causing pain, vesicles
      along the external canal, vertigo and facial nerve palsy

OTITIS MEDIA

Infection of the middle ear cleft, most commonly in infants 6–36
months, more in boys.
   Predisposing factors include: adenoidal masses, allergy,
craniofacial abnormalities, patulous Eustachian tube, immune
defect, disorders of cilial motility.

Diagnostic methods
1. Examination of tympanic membrane with auroscope, including
   assessment of its compliance
2. Aspiration and culture of secretions through tympanic membrane
   (tympanocentesis)
   Culture of nasal or oral secretions is not useful
3. Audiometry

Complications of otitis media
1. Adhesions and ossicle damage, causing hearing loss
2. Cholesteatoma formation
3. Mastoiditis
4. Brain abscess, meningitis, lateral sinus thrombosis

Clinical types of otitis media

a. Acute
— is usually preceeded by viral upper respiratory tract infection causing viral otitis
— bacterial superinfection may follow, due to *Pneumococcus, H. influenzae,* streptococci
— causes earache, fever, hearing loss
— the drum may be red, opaque, bulging. Perforation may follow
— treated with antibiotics, inhalations, decongestants, analgesics. Myringotomy (drainage) may be required

b. Recurrent otitis media
— may be associated with respiratory allergy, adenoid enlargement, rarely IgA deficiency or disorders of cilial motility

c. Serous otitis media ('glue ear')
— cause is uncertain
— treatment consists of trials of antibiotics, myringostomy

d. Chronic suppurative otitis media
— persistent perforation of drum, chronic aural discharge

INFECTIONS RELATED TO TEETH

The normal flora of the 'gingival crevice are most commonly isolated from such infections, including *Bacteroides, Fusobacterium, Peptostreptococcus, Actinomyces, Veillonella* and streptococci.
   Most infections tend to localise and drain spontaneously, occasional infections spread in facial soft tissues or to remote sites
1. Caries (tooth decay)
    — results in the destruction of enamel and dentin
    — due to microbial activity although the precise cause is uncertain
2. Pulp space infection
    — follows destruction of the tooth by caries
    — if drainage is not established results in pulp necrosis, periapical abscess. Perforation of the bony cortex leads to soft tissue infection
3. Periodontitis (pyorrhea)
    — is infection involving the gum, periodontal ligament and alveolar bone around a tooth
    — may progress slowly for many years, usually remains localised to intraoral soft tissues

Sites of spread of odontogenic infections
1. *Fascial spaces*
a. *Submandibular and sublingual spaces*: Ludwig's angina. Causes brawny oedema making swallowing difficult and impairing respiration
b. *Lateral and retropharyngeal spaces*: causes pain, trismus, dysphagia and oedema of the larynx
c. *Masseteric space*
d. *Buccal and parotid spaces*

2. *Maxillary sinusitis*
— follows direct extension of infection into sinus from roots of the maxillary molars

3. *Cavernous sinus thrombosis*

4. *Endocarditis*

Principles of treatment
1. Antibiotics, usually penicillin
2. Drainage

CONJUNCTIVITIS
Infection of the conjunctival sac is characterised by:
1. Redness, conjunctival oedema (chemosis)
2. Exudate. Dried secretions stick the eyelids together after sleep
3. Discomfort
4. Presence of follicles, if prolonged
5. Pre-auricular lymph node enlargement may occur with some infections

Causes
1. *Bacterial*
Pneumococcus
Haemophilus influenzae } Cause most cases
Haemophilus aegyptius } Respond to topical antibiotics
(Koch-Weeks bacillus)
Staphylococcus aureus: associated with chronic infection
Moraxella lacunata (Morax-Axenfeld bacillus) causes angular conjunctivitis with adjacent dermatitis
N. gonorrhoeae: Occasional cause of neonatal conjunctivitis acquired in the birth canal
Chlamydia trachomatis { Trachoma
{ Neonatal conjunctivitis

2. Viral
Herpes simplex: Primary infection causes conjunctivitis, recurrences cause dendritic ulcers
Adenovirus, type 3: accompanied by fever and pharyngitis
Adenoviruses types 8 and 17: acute follicular conjunctivitis followed by keratitis and pre-auricular lymphadenopathy. May be transmitted by instruments
Enterovirus 70: acute haemorrhagic conjunctivitis sometimes associated with paralysis
Vaccinia: Follows accidental splashing of vaccine in the conjunctival sac

KERATITIS
Inflammatory disease of the cornea. Microbial causes include:

1. Bacteria
Staphylococcus aureus
Pneumococcus
Pseudomonas aeruginosa especially in intensive care units

2. Viral
Herpes simplex (dendritic ulcers)
Varicella-zoster, involving the trigeminal nerve
Adenoviruses

3. Fungi
Keratomycosis. Usually with a preceeding history of corneal trauma

Clinical features
1. Pain, diminished visual acuity, limbic inflammation
2. Ulcer formation may lead to vascularisation and scarring
3. Extension of infection to the anterior chamber causes keratitic precipitates, flare, hypopyon

UVEITIS

1. Bacterial endophthalmitis
— usually follows ocular surgery, occasionally trauma.
— causes include:
   Staphylococcus aureus: causes 50% of postoperative infections
   Gram-negative rods, especially *Pseudomonas aeruginosa*
   Fungi

2. Iritis
Inflammatory disease of the anterior uveal tract is usually due to causes other than infectious agents occasional members of which either invade these tissues or cause hypersensitivity.

a. Tuberculosis
b. Syphilis
c. Brucellosis
d. Leprosy
e. Herpes simplex
f. Varicella-zoster
g. Onchocerciasis

3. Chorioretinitis

Infection in the posterior uveal tract follows invasion of the circulation.

a. *Toxoplasmosis*
   accounts for 25–70% of all posterior uveitis and may only be detected years after congenital or acquired infection
b. *Toxocariasis*
   occurs mostly in children after ingesting the ova of these nematodes carried by dogs and cats
c. *Candidiasis*
   follows venous line contamination
e. *Cytomegalovirus*
   is usually congenital, occasionally follows prolonged immunosuppression
f. *Dimorphic fungi*
   Coccidioidomycosis
   Histoplasmosis
g. *Syphilis*
   congenital and acquired
h. *Nocardiosis*
i. *Tuberculosis*
j. *Bacteria from septic foci*
   Staph. aureus
   Strep. pneumoniae
   N. meningitidis

Skin and soft tissue infections

1. IMPETIGO

— Bacterial infection of the epidermis.
— Prepubertal skin is most vulnerable with minor abrasions on exposed areas predisposing.
— Dried purulent discharge creates thick yellow crusts.
— Remains superficial, no scarring occurs.
— There are two agents involved:
  a. *Streptococcus pyogenes*
     — is highly communicable, spreads in families in conditions of poor hygiene
     — is caused by certain M-types, glomerulonephritis may follow
     — requires removal of crusts and systemic penicillin or erythromycin for 10 days
     — may be secondarily colonised by *Staphylococcus aureus*
  b. *Staphylococcus aureus*
     — is due to epidermolysis caused by toxin-producing strains, usually phage type 71
     — causes large flaccid bullae contain clear yellow fluid

2. FOLLICULITIS

— Infection of the hair follicles causing small papules or pustules
— Usually due to *Staphylococcus aureus*, occasionally *Pseudomonas aeruginosa* ('spa pool dermatitis') or *Candida albicans*
— Common on bearded areas (sycosis barbae)

3. FURUNCULOSIS

An inflammatory, later suppurating, nodule of the subcutaneous tissue developing from folliculitis and nearly always due to *Staphylococcus aureus*.

Predisposing factors include diabetes mellitus, defects in neutrophil function, steroid therapy or blood dyscrasias. But in most cases, especially those which occur in families, invasive strains of *Staph. aureus* are involved, the pathogenicity of which is not yet explained.

Management includes:
— drainage when fluctuant
— antibiotics: cloxacillin, erythromycin, tetracycline
— intranasal antibiotics (bacitracin, neomycin)
— antiseptic emulsions (hexachlorophane, chlorhexidine)
— colonisation by interfering strains of *Staph. aureus*

4. ACNE

— The primary disorder is retention of sebaceous secretions during and after puberty, creating comedones ("blackheads").
— Followed by colonisation and infection with normal skin flora, especially Propionibacterium acnes.
— Prolonged antibiotic administration (tetracyclines, erythromycin) reduces incidence of pustule formation and scarring.

5. SUPPURATIVE HIDRADENITIS

Infection of apocrine sweat glands usually by mixed skin flora with anaerobes predominating.

Frequent recurrences may necessitate excision and grafting.

6. CELLULITIS

— An acute spreading infection of the subcutaneous tissues.
— May be associated with lymphadenitis or lymphadenopathy
— Causes tenderness, pain, swelling, erythema.
— Has several causes, including:
   a. *Streptococcus pyogenes* (occasionally other beta-haemolytic streptococci as well)
      — causes a distinctive superficial cellulitis (erysipelas) in younger children and older adults, especially involving the face
      — characterised by bright red induration, advancing raised border, together with fever
      — portal of entry cannot usually be found and the organism cannot be cultured from the overlying skin
      — may follow trauma, surgical wounds, varicose ulcers; rapidly invading the blood stream or causing tissue necrosis (streptococcal gangrene)

b. *Staphylococcus aureus*
   — is the most frequent cause of cellulitis following trauma or clean surgery, especially in the presence of foreign bodies
   — causes purulent discharge, occasionally fever
   — management includes: wound lavage with saline or hypochlorite, suture removal, exploration for foreign bodies, antibiotics for constitutional symptoms
c. Anaerobes
   — usually associated with aerobes, especially in sites of contamination with bowel organisms (perianal, abdominal wounds after surgery)
   — may cause erythema only or progress to extensive necrosis of subcutaneous tissues (gangrene) with gas formation, involvement of deep fascia and muscles
   — predisposing factors include: devascularisation, foreign body, carcinoma, malnutrition, alcoholism
   — associated with several clinical syndromes:
     (i) Gas gangrene (clostridial cellulitis and myonecrosis) due to toxin formation
    (ii) Meleney's synergistic gangrene; occurs at abdominal wound sites, due to *Staph. aureus* plus microaerophilic or anaerobic streptococci
   (iii) Necrotising fasciitis

7. INTERTRIGO

Infection of moist areas of skin (groins, axillae, corners of mouth, beneath pendulous breasts).
   Mixed organisms, especially *Candida albicans*.

8. SECONDARY BACTERIAL INFECTION OF PRE-EXISTING SKIN LESIONS

a. **Burns**
   — are prone to colonisation of the eschar, usually by the 10th day
   — Gram-negative bacteria, especially *Pseudomonas aeruginosa*, predominate. Occasionally staphylococci, streptococci, fungus, herpes simplex virus
   — the most common cause of death, especially when total burn area exceeds 60%
   — sepsis is frequent when eschar bacterial count exceeds 10^5 organisms/g but blood cultures are usually negative
   — management includes:
     (i) local antibacterials: silver sulphadiazine, sulfamylon
    (ii) during serious sepsis: antipseudomonal antibiotics
   (iii) quantitative bacterial counts of excised eschar may help decide antibiotic management

b. Skin ulcers (varicose, bed sores)
— become colonised with mixtures of skin flora, coliforms and
  anaerobes
— may be complicated by cellulitis (esp. due to *Streptococcus
  pyogenes*) or bacteraemia

c. Surgical wounds (see chapter on Hospital Infections)

**9. CUTANEOUS INFECTIONS due to agents aquired by occupation
   or in certain geographical areas**

— may be zoonotic with man as an incidental host
— usually involve extremities or exposed areas
— are often preceeded by trauma
— may be due to organisms which look nonedescript on usual
  laboratory media

a. Anthrax
— is due to Bacillus anthracis: the environment is the major
  reservoir and the disease occurs primarily in herbivores
— results in:
    (i) necrotic spreading lesions on the extremities
    (ii) pneumonia, if the spores are inhaled
— is managed with penicillin, patient isolation, quaranting of herds
— vaccines are effective

b. Erysipeloid
— is caused by *Erysipelothrix rhusiopathiae*, a commensal of
  animals, birds and fish
— most common in workers with pigs and fish
— is characterised by slowly progressive painful swelling around
  the site of inoculation (erysipeloid of Rosenbach)
— occasionally causes bacteraemia, endocarditis
— responds to penicillin

c. Glanders
— is a subcutaneous infection of equines caused by *Pseudomonas
  mallei*, occasionally transmitted to man

d. Melioidosis
— is caused by *Pseudomonas pseudomallei*, an environmental
  organism, most common in tropical countries especially in
  Southeast Asia
— causes nodules, sinuses, lymphadenopathy, septicaemia,
  pneumonia, occasionally months or years after leaving the area
  of acquisition
— is managed with chloramphenicol, tetracycline, trimethoprim
  compound

e. Vibrios
— occasionally cause infection of wounds sustained in brackish or salt water

f. Animal and human bites:
— usually mixed infections which may include
  (i) Anaerobes
  (ii) *Streptococcus pyogenes*
  (iii) *Streptococcus milleri*
  (iv) *Pasteurella multocida* (cats and dogs)
  (v) *Eikenella corrodens*

g. Mycobacterial infections (Table 80)

Table 80

Species	Habitat	Clinical features	Treatment
M.marinum	Swimming pools Fish tanks	Infects abrasions Causes ulceration and lymphadenitis	Slow spontaneous healing Rifampicin, ethambutol
M.ulcerans	Environment of tropical areas	Buruli or Bairnsdale ulcer can cause extensive tissue destruction	Resistant to drugs Wide excision and grafting, heat cradles
M.fortiutum *M.chelonei*	Environment (worldwide)	Follows penetrating or surgical wounds	Tetracyclines, sulphonamides Debridement

h. Animal and human poxviruses
Vaccinia (cowpox): Used in vaccination against smallpox for nearly two centuries
Paravaccinia: Disease of cattle, causes milkers' nodules
Orf: Causes contagious pustular dermatitis of sheep and goats, causes nodular lesions on the hands of farmers
Molluscum contagiosum: Sexually transmitted disease of man

i. Sporotrichosis
Due to an environmental fungus, *Sporothrix schenckii*; causes cutaneous nodules and ulcers with subsequent lymphangitis.

j. Mycetoma (madura foot)
— caused by saprophytic soil fungi and aerobic actinomycetes
— most frequent in the tropics, commonly on the feet
— causes suppurative granuloma, containing coloured mycotic
  grains which discharge through draining sinuses
— progressive suppuration, fistula and scar formation create a
  painless, massively swollen and deformed foot
— managed with débridement, often amputation. Antimicrobials
  are ineffective

k. Leishmaniasis
— is due to intracellular protozoans transmitted by the bite of
  sandflies

10. SKIN LESIONS DUE TO CIRCULATING TOXINS (Table 81)

Table 81

Disease	Agent	Toxin
Scarlet fever	Streptococcus pyogenes	Erythrogenic toxin
Toxic shock syndrome	Staphylococcus aureus	? Enterotoxin F
Lyell's syndrome Ritter's disease	Staphylococcus aureus bacteriophage type 71	Dermatotoxin
Kawasaki disease	Cause unknown	Possibly toxic

11. HAEMATOGENOUS INFECTIONS

Skin lesions which follow spread of infectious agents within the
circulation include:
Neisseria: Vasculitic lesions due to deposition of meningococci or
  gonococci
Endocarditis: a. immune complex disease
                   b. embolic lesions
Necrotic lesions in neutropenic hosts: Pseudomonas aeruginosa
  (ecthyma gangrenosum), staphylococcal and fungal infections
Viral exanthemata: Measles, rubella, varicella, enteroviruses,
  infectious mononucleosis
Rickettsial infections: Typhus, scrub typhus
Rose spots of typhoid
Viral haemorrhagic fevers, arboviruses: see chapter on fever

LYMPHADENITIS

Acute or chronic inflammation of lymph nodes

1. Regional (Table 82)

Table 82

Common bacteria of skin and mucous membranes	Streptococcal	Strep. pyogenes
	Staphylococcal	Staph. aureus
	Anaerobic	Anaerobes
Zoonoses, following insect bites	Tularaemia	F. tularensis
	Plague	Y. pestis
	Cat scratch disease	unidentified bacterium
	Rat-bite fever	S. moniliformis
		S. minus
	Typhus	Rickettsiae
Environment	Sporotrichosis	S. schenckii
	Glanders	P. mallei
	Melioidosis	P. pseudomallei
	Anthrax	B. anthracis
Mycobacteria	Scrofula	M. tuberculosis
		M. scrofulaceum
	Swimming pool granuloma	M. marinum
Venereal diseases	Syphilis	T. pallidum
	Herpes	Herpes simplex
	Lymphogranuloma venereum	C. trachomatis
	Granuloma inguinale	C. granulomatis
	Chancroid	H. ducreyi

2. Generalised

Acute generalised lymphadenitis is usually due to viruses rather than bacteria. The nodes are typically tender, discrete and firm.

Prolonged lymphadenopathy has several important causes of which infections are only one of several possibilities

a. Infections
   — toxoplasmosis, syphilis, histoplasmosis, acquired immune deficiency syndrome.
b. Neoplasia
   — lymphoma, leukaemia
c. Hypersensitivity
   — hydantoins, serum sickness
d. Connective tissue diseases
   — rheumatoid arthritis, systemic lupus erythematosis

e. Endocrine
— hyperthyroidism, Addison's disease
f. Storage disease
— Gaucher's, Niemann-Pick's diseases
The infectious causes of generalised lymphadenopathy are best classified according to the class of microbe responsible:

Viral
— Infectious mononucleosis
— Measles, Rubella
— Adenoviruses
— Acquired immunodeficiency syndrome (AIDS)
— Haemorrhagic fevers
Bacterial
— Syphilis
— Miliary tuberculosis
— Leptospirosis
— Brucellosis
— Scrub typhus
Fungal
— Histoplasmosis
Protozoan
— Toxoplasmosis
— Leishmaniasis
— Trypanosomiasis
Metazoan
— Filariasis

Infections of the female genital tract

VAGINITIS

Inflammation of the vagina presenting with vaginal discharge and/or vulval pruritus.

Causes

1. *Candidiasis*
   — is caused by *Candida* species, normal vaginal flora whose overgrowth is encouraged by:
   a. hormonal change (pregnancy, oral contraceptives, premenstrual state)
   b. antibiotics
   c. diabetes mellitus
   — presents with pruritis and relatively scanty discharge. The introitus is reddened, vaginal wall may have 'cheesy' deposits
   — responds to topical applications of polyenes or antifungal imidazoles. Relapse is frequent if treated for less than 10 days

2. *Trichomoniasis*
   — is caused by the flagellate protozoan *Trichomonas vaginalis*, acquired by sexual contact
   — becomes symptomatic at menstruation when vaginal secretions and increasing pH encourage multiplication of the organism
   — causes symptoms which range from nil to profuse frothy discharge and irritation. Males are usually symptomless.
   — responds to treatment with antiprotozoan imidazoles. Concurrent treatment of sexual partner reduces the risk of relapse

3. *Gonorrhoea* (see next chapter)

4. 'Non-specific'
— is characterised by thin, grey, malodorous discharge which contains few polymorphs (i.e. vaginosis rather than vaginitis)
— discharge has a pH of 5.0 or greater and contains amines
— culture reveals predominant *Gardnerella* (previously *Haemophilus*) *vaginalis* and anaerobes, in the absence of lactobacilli
— responds to treatment with antianaerobe imidazoles

5. Atrophic (senile)
— occurs after the menopause when mixed aerobes, rather than lactobacilli, dominate the vaginal flora
— responds to topical oestrogens

URETHRITIS

Dysuria and frequency in the absence of bladder infection ('urethral syndrome') occurs commonly in women yet few causes of urethritis have been elucidated. They include:
1. Gonorrhoea (see next chapter)
2. *Chlamydia trachomatis*
3. Associated with cystitis or vaginitis

CERVICITIS

Infection of the columnar epithelium of the endocervix produces mucoid or purulent cervical secretions which are usually insufficient to be recognised as vaginal discharge. Causes include:
1. Gonorrhoea
2. *Chlamydia trachomatis*
3. Herpes simplex

INVESTIGATION OF VAGINAL DISCHARGE

Swabs should be taken from the vaginal vault and endocervix, as well as the urethra if there is urethral discharge.

1. Microscopy
a. *Wet film*: immediate examination reveals the motile trophozoites of *Trichomonas*. Yeasts may also be seen
b. *Stained film*: Gram stain reveals yeasts and bacteria as well as polymorphs. Papanicolaou stain (preferred for cytology) may also reveal *Trichomonas*
c. *Immunofluorescence*: Monoclonal IFA techniques have recently been devised for *Chlamydia*

2. Culture
a. *Blood agar*: useful for Gardnerella, coliforms and yeasts but may be misleading
b. *Gonococcus* requires the use of selective media (see chapter on gram-negative cocci)
c. *Yeasts*: Selective media (such as Sabouraud's) can be used but are usually unnecessary
d. *Trichomonas*: Culture in a selective liquid medium is a slower but somewhat more sensitive method of detection than microscopy
e. *Chlamydia* requires tissue culture techniques

3. Detection of antigens
An ELISA test is available for the detection of antigens of *N.gonorrhoeae* and *C.trachomatis*

PRINCIPLES OF MANAGEMENT OF VAGINAL DISCHARGE:

1. Complete gynaecological history and pelvic examination are important
2. Specimens must be appropriately collected and examined
3. Predisposing factors (diabetes, oestrogen deficiency) must be sought
4. Other causes of vaginal discharge (malignancy, chemicals, foreign body, exaggerated physiological secretion) should be considered
5. Patient compliance and treatment of sexual partners reduce the risk of relapse

SALPINGITIS

— Infections of the fallopian tubes. The term 'pelvic inflammatory disease' is used interchangeably
— Results from ascending infection, unrelated to surgery or pregnancy
— Organisms spread from the endocervix through the endometrium to the fallopian tubes, especially during or after menstruation
— Unless treated early, destruction of the cilated epithelium results in partial or complete tubal occlusion
— Consequences include:
  1. Hydrosalpinx, pyosalpinx
  2. Tubo-ovarian abscess — when infected material penetrates ovarian tissue
  3. Pelvic peritonitis, which may spread to the upper peritoneum (perihepatitis, Fitz-Hugh–Curtis syndrome)
  4. Recurrent episodes of salpingitis
  5. Ectopic pregnancy
  6. Infertility

Causes

1. Neisseria gonorrhoeae
— causes up to 90% of first episodes
— predisposes to recurrent non-gonococcal infection

2. Chlamydia trachomatis
— has recently been established as a cause of salpingitis the incidence of which probably exceeds that of gonorrhoea in many population groups

3. Anaerobes (normal vaginal flora)
— seem to play an important role in recurrent infection

Clinical features
1. Fever, chills, malaise
2. Bilateral lower abdominal pain, causes patient to crouch when standing (due to psoas irritation)
3. Most often developes during or immediately after menstruation
4. Cervical discharge
5. Bimanual pelvic examination reveals uterine and adnexal tenderness, palpable pelvic 'thickening' or a tubo-ovarian mass
6. Right upper quadrant pain (perihepatitis) if peritoneal spread occurs
7. Differential diagnosis includes septic abortion, appendicitis, ectopic pregnancy, torsion or rupture of an ovarian cyst, endometriosis and pyelonephritis

Diagnostic methods
— Include:
   1. Endocervical culture
   2. Culdocentesis
   2. Pelvic ultrasound
   3. Laparoscopy

Treatment
1. Should include an antibiotic effective against *N.gonorrhoeae* (penicillin, amoxycillin)
2. May also include antibiotics for anaerobes or *C.trachomatis* (metronidazole, tetracyclines)
3. Surgery is indicated for
   a   pelvic abscess not responding to treatment
   b. doubt about the diagnosis
   Chronic pelvic inflammatory disease includes the following entities:
1. Recurrent salpingitis following gonococcal or possibly chlamydial infection
2. Tuberculous salpingitis
3. Pelvic actinomycosis following intrauterine contraceptive device

ENDOMETRITIS

Infection of the uterine cavity following abortion, childbirth or surgery.
Predisposing factors:
1. Caesarian section
2. Factors associated with difficult deliveries, viz.
   a. prolonged labour and rupture of membranes
   b. repeated pelvic examination during labour
   c. episiotomy, use of forceps, trauma, haemorrhage, anaemia
3. Retained products of conception
4. Illegal or inexpert manipulations
   The organisms involved are all members of the normal vaginal flora, usually in mixed culture:

Anaerobic cocci and
Bacteroides species
— The most frequent cause of endometritis with insidious onset of fever which usually resolves spontaneously
— Occasionally leads to high swinging fever and septic pelvic thrombophlebitis

Bacteroides fragilis
— less common, more likely to cause prolonged infection which does not respond to penicillin

Clostridium welchii
— occasional cause of overwhelming sepsis following septic abortion

Streptococci
— gp B are frequently isolated
— gp A, previously the most important cause of puerperal sepsis is now rare but still potentially lethal

Mycoplasmas
— may rarely cause prolonged swinging fever without localising signs or toxicity
— Responds to tetracycline

Coliforms
— May cause early gram negative septicaemia from urinary tract as well as endometrium

Staph. aureus
— is an unusual cause of endometritis. Associated with cervical ligature, diabetes, episiotomy. Early sepsis without shock

Listeria monocytogenes and
Campylobacter fetus
— Rare causes of infection during and after childbirth

Diagnostic methods
1. Vaginal culture will usually yield the pathogen but often as part of a general colonisation of the lochia with faecal flora. May be misleading
2. Blood culture yields pathogens in about 8% of cases, more frequently if sepsis is severe
3. Pelvic ultrasound may be indicated if fever is prolonged

Principles of management
1. Initial clinical assessment should consider all the causes of puerperal fever, viz.
   a. Endometritis
   b. Renal tract infection
   c. Surgical wound (caesarian section) infection
   d. Pneumonia
   e. Mastitis
   f. Deep venous thrombosis
   g. Drug fever
   h. Sepsis from venous lines
   i. Infection from blood transfusion, especially cytomegalovirus
2. Appropriate specimens should be collected (vagina, urine, blood, breast milk)
3. Antibiotic therapy should be commenced according to the clinical state, viz.
   a mild: oral penicillin, amoxycillin or metronidazole
   b. severe: intravenous combinations of penicillin, aminoglycoside and metronidazole
4. If fever persists, pelvic examination, ultrasound, chest X-ray, full blood count are indicated to exclude extension of infection and to eliminate other causes of fever

CHORIOAMNIONITIS

Acute inflammation of the fetal membranes:
1. Is present in up to 20% of live births
2. Increases in incidence and severity with prolongation of labour
3. May be the cause rather than the result of premature labour (between 20 and 28 weeks)
4. Represents an ascending infection through the cervix, involving the amniotic fluid, chorionic plate, membranes, umbilicus and eventually, the fetal lung
5. Is caused by normal vaginal flora, including group B streptococci, anaerobes, Esch. coli and mycoplasmas
6. Produces mild fever and leucocytosis in the mother
7. May be treated with the same antibiotics as used for severe endometritis but early delivery is indicated

NEONATAL INFECTIONS

Infections in the first days of life have several origins.

Transplacental infections
— are recalled by the acronym TORCH (*t*oxoplasmosis, *o*thers, *r*ubella, *c*ytomegalovirus, *h*erpes simplex). A more complete list is as follows:
  1. Viruses
     a. Rubella
     b. Cytomegalovirus
     c. Herpes simplex
     d. Varicella
  2. Bacteria
     a. Syphilis
     b. Listeriosis
     c. Campylobacter fetus
  3. Protozoans
     a. Toxoplasmosis
— The outcome of transplacental infection is related to the time of infection, viz.
  a. 1st–2nd months: spontaneous abortion or multiple defects
  b. 3rd–4th months: moderate risk of single defects
  c. 5th–6th months: slight risk of single defects
— The clinical syndrome includes
  a. Early transient signs
     — Low birth weight
     — Thrombocytopenic purpura
     — Hepatosplenomegaly
     — Lymphadenopathy
     — Pneumonitis
  b. Permanent
     — Deafness
     — Cataract
     — Congenital heart disease
     — Mental retardation
     — Chorioretinitis

Birth canal
Infections acquired in the birth canal or ascending into the amniotic fluid affect newborns as follows:
a. *Strep. agalactiae* (group b streptococcus)
   Pneumonia
   Meningitis
b. *N.gonorrhoeae* Conjunctivitis
c. *Chlamydia trachomatis*
   — Conjunctivitis
   — Pneumonia

d. Herpes simplex
   — Neonatal herpes, often disseminated
e. Hepatitis B
   — Hepatitis with lifelong antigenaemia and risk of hepatoma
f. Candidiasis
   — Neonatal thrush
g. Listeria
   — Listeriosis
h. Warts
   — Laryngeal papillomas
i. Cytomegalovirus
   — Neonatal CMV (usually asymptomatic)
j. Mixed flora (see chorioamnionitis)
   — Aspiration pneumonia
k. Enteric pathogens
   — *Salmonella*, *Shigella* and *Campylobacter* may occasionally be
     acquired in the birth canal

Acquired in the neonatal ward
These infections come from other babies, from the attendants, from
the mother, her visitors or from the environment.
a. *Staphylococcus aureus*
   A common cause of nursery epidemics spread by hands and first
   colonising the umbilical stump
b. *Enteric organisms*
   Enteropathogenic *Esch. coli*
   Rotaviruses
   Enteroviruses
c. *Respiratory viruses*
d. *Herpes simplex*
e. *Environmental organisms*
   Pseudomonas aeruginosa and coliforms may sometimes
   contaminate infant feeding formula or nebulisers causing
   invasive disease in premature neonates
f. *Tetanus*
   follows unhygienic methods for cutting or covering the cord

Sexually transmitted diseases

Although numerous infections may be transmitted by sexual contact, the term is usually kept for infections causing genital lesions (see Table 83).

1. Ulcers (see table)
Herpes simplex: Herpes simplex virus
Syphilis: *Treponema pallidum*
Chancroid: *Haemophilus ducreyi*
Lymphogranuloma venereum: *Chlamydia trachomatis*, serotypes L1, L2, L3
Granuloma inguinale (Donovanosis): *Calymmatobacterium granulomatis*

2. Discharge
Gonorrhoea: *Neisseria gonorrhoeae*
Non-gonococcal urethritis (males): *Chlamydia trachomatis*, serotypes A-K
Trichomoniasis (females): *Trichomonas vaginalis*

3. Papules
Warts: Papillomavirus
Molluscum contagiosum: (Unclassified poxvirus)

3. Infestations
Scabies: *Sarcoptes scabiei*, var. *hominis*
Pediculosis ('crabs') *Phthirus pubis*

SYPHILIS

A systemic illness caused by the spirochaete *Treponema pallidum*, usually transmitted by sexual contact.

The prolonged natural course of the disease is divided into several stages, viz,

Table 83 Sexually transmitted diseases causing genital ulceration

Disease	Causative agent	Incubation period in days	Local symptoms	Diagnostic methods	Systemic manifestations	Treatment
Syphilis	*Treponema pallidum*	21 (9–90)	Single painless ulcer with raised firm borders. Moderately large, painless, non-suppurating local lymph nodes.	Dark ground examination of ulcer Serology	Secondary and tertiary disease	Penicillin Tetracycline Erythromycin
Herpes simplex	Herpes simplex virus (usually type II)	6 (2–21)	Painful vesicular lesions which rapidly ulcerate. Tender lymph nodes. Recurrent ulcers are smaller and heal in 8–10 days	Tzanck preparation Electron microscopy Virus culture (serology is usually not helpful)	Fever and meningism in primary infection	Vidarabine Acyclovir
Chancroid	*Haemophilus ducreyi*	3 (1–4)	Sharply demarcated painful ulcer. Superinfection may follow. Painful lymphadenopathy, often suppurating	Demonstration of the organism by smear or culture of lesions	Nil	Sulphonamides Tetracycline Streptomycin Trimethoprim

	Organism	Incubation (days)	Clinical features	Diagnosis	Complications	Treatment
Lympho-granuloma venereum	*Chlamydia trachomatis* serotypes L1, L2, L3	(3–21)	Transient papule or ulcer followed by suppurative lymphaadenopathy, fistula formation, infiltration and fibrosis	Culture Serology (CFT, IFA)	Fever, myalgia Aseptic meningitis Arthritis Erythema nodosum Hepatitis	Sulphonamides Tetracycline
Granuloma inguinale (Donovanosis)	*Calymmato-bacterium granulomatis*	(8–80)	Heaped granulomatous ulcer, extending and fibrosing. Lymphatics only occasionally involved.	Donovan bodies seen in Giemsa-stained scrapings of granulation tissue. Organism cannot be cultured	Occasional metastatic spread to bones, joints, liver	Tetracycline Streptomycin Gentamicin Chloramphenicol

1. Primary (the chancre)
— incubation period about 3 weeks
— initially a nodule on penis, labia, cervix, anus, rectum or lips, soon becomes a painless ulcer which heals in 2–3 weeks
— accompanied by painless, non-suppurating regional lymphadenopathy

2. Secondary
— soon follows the healing of the chancre, causing:
  a. flu-like illness: fever, arthralgia, headaches, weight-loss
  b. generalised lymphadenopathy, splenomegaly
  c. skin lesions: generalised, especially palms and soles, non-itching, discrete, pink to red. In moist areas may form broad plaques (condyloma lata)
  d. mucosal lesions: mucous patches (snail track ulcers)
  e. organ involvement: meningitis, glomerulonephritis, hepatitis, uveitis, arthritis

3. Latent
Asymptomatic, but relapses of secondary lesions can occur for up to 4 years after onset.

4. Tertiary (late)
10–40 years after onset, in 30–50% of untreated patients

a. Central nervous system
 (i) meningovascular
(ii) parenchymatous:
   General paresis: dementia, delusions, Argyll–Robertson pupils, hyperreflexia, optic atrophy
   Tabes dorsalis: ataxia, shooting pains, sensory loss, Rhombergism, areflexia

b. Cardiovascular
Endarteritis of ascending aorta causing aortic aneurysm, coronary stenosis.

c. Granulomatous lesions (gummata), which can occur in any organ but in particular:
 (i) Bones
 (ii) Skin: nodules, punched-out ulcers
(iii) Mucosa

5. Congenital
Transplacental infection can occur even in latent syphilis but usually in the first year.
 Characteristic features include:
Early
— Osteochondritis
— Rhinitis ('Snuffles')
— Desquamating maculopapular rash
— Hepatosplenomegaly
Late stigmata
— Saddle nose
— Interstitial keratitis
— Knee effusions (Clutton's joints)
— Peg-shaped (Hutchinson's) teeth

Diagnosis
1. Demonstration of motile spirochaetes in exudates from chancre or mucosal lesions examined by dark field microscopy.
2. Serology (see chapter on Spirochaetes etc)

Treatment
1. Primary and secondary
    — Benzathine penicillin G, 2.4 million units imi in a single dose or
    — Procaine penicillin 600 000 U daily i.m. for 8–10 days.
2. Latent and tertiary
    — Procaine penicillin 600 000 U daily i.m. for 14 days.
    — Probenicid is often added

Herpes simplex (see also chapter on DNA viruses)
Although both types of herpes simplex virus (HSV) can cause genital ulceration, type II accounts for 80–90% and is more likely to cause recurrence.

Clinical features
1. Primary
In young adults who have had no prior exposure to HSV (e.g. labial herpes), the primary attack is severe with:
a. Vulva: Intense oedema, extensive painful ulcers. Acute retention may follow
b. Penis: Extensive vesiculation
c. Rectal: Tenesmus, discharge, rectal bleeding
 Other manifestations in all these groups include:
a. inguinal lymphadenopathy
b. systemic symptoms (malaise, fever)
c. autoinoculation lesions (pharynx, thighs, hands, face)
d. meningism

2. Recurrent
— is due to HSV II in more than 90%
— tends to diminish over time
— causes lesions which vary in size and duration from one episode to the next but on average, shed HSV for 5 days and heal in 8–10 days
— attacks are precipitated by menses, trauma during intercourse, stress and immunosuppression

Diagnosis
See chapter on DNA Viruses

Treatment of genital herpes
1. Primary
— Intravenous vidarabine or acyclovir reduce the duration of symptoms but do not prevent recurrences
2. Recurrent
— Topical acyclovir has little effect on the duration of symptoms
— Oral acyclovir prevents recurrences only while the drug is administered
— Numerous other therapies have not been proven to be effective in clinical trial

GONORRHOEA

N.gonorrhoeae adheres to columnar and transitional epithelium causing a purulent exudate by mechanisms which are not well understood.

Ascending and invasive disease is much more frequent in females despite less clinically apparent disease (Table 84)

Table 84

	Males	*Females*
Site of primary infection	Anterior urethra	Endocervix
Asymptomatic disease	5%	30–50%
Genital symptoms	Purulent urethral discharge Dysuria	Dysuria, discharge (often mild)
Ascending infection	Epididymitis Orchitis Prostatitis	Salpingitis (10–20%) Tubal abscess Sterility

Disseminated gonococcal infection
— occurs in 1–3% of, usually asymptomatic, patients
— occurs more frequently in females than males, usually at the onset of menstruation
— is occasionally associated with inborn defects of complement production, more commonly with certain types of gonococci
— causes fever, polyarthralgia, tenosynovitis and a rash characterised by discrete petechial or pustular lesions on the extremities
— may cause frank purulent arthritis (the commonest cause in young adults)

Diagnosis of gonorrhoea (Table 85)

Table 85

	Males	*Females*
Method of specimen collection	Swab of pus from urethral meatus or wire loop placed in anterior urethra	Cervix must be visualised and swab placed in endocervix
Gram stain	Reveals numerous intracellular Gram-negative diplococci	Either negative or difficult to interpret
Culture	Yields profuse *N.gonorrhoeae.*	Must be made on selective media to prevent growth of normal flora
Other sites	Rectum and pharynx in male homosexuals	Urethra, rectum, pharynx

Antibiotics for gonorrhoea
— Penicillins are still the antibiotics of choice and give cure rates, even with single doses, of about 95%
— Alternative treatment is necessary if
　a. the patient is allergic to penicillin.
　b. the strain of *N.gonorrhoeae* produces beta lactamase
— Cure should always be confirmed by follow-up and in the case of females, by repeated cervical culture

Benzyl penicillin: used intravenously for systemic infection

Procaine penicillin: 2.5–4.0 Mu i.m. together with probenicid 1.0 g p.o., as a single dose

Ampicillin ⎫ 3.0–3.5 g p.o. plus probenicid 1.0 g p.o., as a single
Amoxycillin ⎬ dose

Tetracyline: 1.0 g p.o. followed by 0.5 g 6-hourly for 5 days

Spectinomycin ⎫
Kanamycin ⎪ Alternative antibiotics for patients with infection
Cefuroxime ⎬ due to beta lactamase-producing *N.gonorrhoeae*
Cefoxitin ⎭

NON-GONOCOCCAL URETHRITIS (NGU)

— Infection of the male urethra due to organisms other than
  N.gonorrhoeae. May be acquired at the same time as gonorrhoea
— About half are caused by *Chlamydia trachomatis*. Some of the
  remainder are due to *Ureaplasma urealyticum*, the cause of the
  rest is unknown
— Occasional complications of NGU include epididymo-orchitis,
  prostatitis and Reiter's syndrome
— The differences between gonorrhoea and NGU in males are
  listed in Table 86

Table 86

	Gonorrhoea	Non-gonococcal urethritis
Incubation period	2–4 days	7–14 days
Onset	Abrupt	Insidious
Discharge	Copious, purulent	Moderate, greyish
Diagnosis	Gram stain and culture	Usually by exclusion
Treatment	Penicillins Tetracycline	Tetracycline Erythromycin

It is important to note that there is sufficient overlap between the
symptoms of both diseases to make laboratory diagnosis necessary.

PRINCIPLES OF MANAGEMENT OF SEXUALLY TRANSMITTED DISEASES

1. Treatment, especially of gonorrhoea and syphilis, is usually
   aimed to render the patient non-infective after a single dose of
   antibiotic
2. Sexual contacts should be traced and investigated or treated
   empirically
3. A serological test for syphilis should be performed on all patients
   with sexually transmitted diseases
4. It is important to see the patient after treatment to verify its
   efficacy

Septicaemia

A clinical syndrome characterised by signs of overwhelming infection including haemodynamic change.

PATHOGENESIS

1. Invasion of the circulation by bacteria or their products, especially endotoxins and teichoic acids
2. Platelets and neutrophils marginate in the vascular bed, causing thrombocytopenia and neutropenia
3. Release of kinins. Activation of coagulation and complement cascades, consumption of clotting factors
4. Peripheral vasodilatation, capillary leakage, lowered venous filling pressure, hypotension
5. Pulmonary changes are prominent, viz.
   a. arterio-venous shunting causes hypoxia and metabolic acidosis.
   b. pulmonary vascular damage causes oedema ('shock lung', adult respiratory distress syndrome)
6. Terminal ('cold') shock results from:
   a. peripheral vasoconstriction in response to diminished circulating blood volume and sympathomimetic activity
   b. hypoxia, metabolic acidosis
   c. diminished myocardial contractility

CLINICAL FEATURES

1. Presence of predisposing factors (see below)
2. Fever, rigors
3. Hypoxia, dyspnoea, tachypnoea, cyanosis
4. Tachycardia, hypotension, congestive cardiac failure
5. Confusion
6. Oliguria
7. Bleeding

COMMON PORTALS OF ENTRY

1. Urinary tract
2. Sites of venous access
3. Peritoneal cavity
4. Lung (pneumonia)
5. Surgical wounds
   Post-partum uterus
6. Bowel wall (in neutropenic patients)

PREDISPOSING FACTORS

1. Instrumentation
   — urinary catheter
   — sites of venous access
2. Sites of gross sepsis
   — peritonitis
   — endometritis
   — infected burns
   — major trauma
3. Immune defect
   — alcoholism
   — uncontrolled diabetes
   — uraemia, dialysis
   — old age
   — prematurity
   — splenectomy
   — cirrhosis
   — neutropenia
   — immunosuppression, steroid therapy

DIAGNOSIS

1. Culture of blood and other sites.
   (bacteraemia is detected in 50–70% of cases).
2. Haematological change
   — neutropenia, later neutrophilia
   — toxic changes in neutrophil morphology
   — increased fibrinolysis, thrombocytopenia
3. Cardiopulmonary changes
   — hypoxia, metabolic acidosis
   — hypotension, decreased cardiac output

Methods of blood culture:
1. Collection of 20–40 ml of blood, usually in aliquots of 5 ml each placed in 45 ml of liquid culture medium
2. Constituents of medium
   a. peptones
   b. supplements
      (i) polyanethanol sulphonate (Liquoid) — prevents clotting and neutrophil phagocytosis
      (ii) reducing agents — for anaerobes
      (iii) antibiotic antagonists — PABA, penicillinase, resins
      (iv) osmolality — may be raised (by adding sucrose) to encourage the growth of cell wall deficient organisms
3. Detection of bacterial growth
   — turbidity. Takes 15–18 hours in most cases
   — Gram stain
   — blind subculture to solid media, in first 24 hours
   — evolution of $^{14}CO_2$ when glucose ^{14}C is present in the medium (Bactec method)
4. Skin contaminants may be present if blood is inexpertly collected. Common species include:
   a. *Staphylococcus epidermidis*, occasionally *Staph. aureus*
   b. *Corynebacterium* species ('diphtheroids')
   c. *Proprionibacterium* acnes ('anaerobic diptheroids')
   d. *Bacillus* species

PRINCIPLES OF MANAGEMENT OF SEPTICAEMIA

1. Collect blood, urine and material from other appropriate sites, for culture
2. Give appropriate antibiotics by intravenous route
3. Expand plasma volume
4. Give oxygen
5. Steroids may be used in large doses at onset. Efficacy is still debated
6. Monitor vital signs and relevant laboratory parameters
   a. central venous pressure
   b. left atrial pressure (Swan–Ganz catheter)
   c. arterial pressure
   d. blood gases
   e. electrolytes
   f. blood glucose
   g. haemoglobin, white cells, platelets, coagulation
   h. chest X-ray
   i. antibiotic levels
7. Remove sources of infection
8. Vasopressors — may be necessary for persistent hypotension
9. Inotropic agents may have a role

10. Experimental therapies include:
    a. Endorphin anatagonists (naloxone)
    b. Arachidonic acid antagonists (indomethacin)
    c. Antibodies to gram negative common antigen
    d. Fibronectin
    e. Heparin
    f. Alpha adrenergic blockers
    g. Glucagon

Principles of antibiotic therapy

1. Use maximal doses of bactericidal antibiotics
2. Use antibiotics appropriate for the clinical situation, or combinations if the cause is uncertain
3. Give intravenously
4. Do not mix two drugs in the same solution or giving set
5. Stop unnecessary drugs when the cause is known
6. Avoid toxicity. Monitor aminoglycoside levels

Intra-abdominal infections

PERITONITIS

Infection of the peritoneal cavity has four sources:
1. 'Primary' infection
2. Secondary to viscus perforation
3. Ascending through fallopian tubes (salpingitis, see chapter on Infections of the Female Genital Tract)
4. Introduced during peritoneal dialysis

Primary peritonitis
— implies peritonitis without an evident source
— is presumed to have come from the circulation or possibly from migration of organisms through the intact bowel wall
— occurs particularly in
  a. nephrotic syndrome
  b. cirrhosis with ascites
— is caused by
  a. Streptococci: *Strep. pneumoniae, Strep. pyogenes*
  b. Enteric organisms: *Esch. coli*, coliforms
— is occasionally caused by *M.tuberculosis* following haematogenous dissemination from the lung

Peritonitis secondary to viscus perforation:
Common origins of peritoneal soiling include:

Appendix:	perforation
Stomach:	perforated peptic ulcer
	carcinoma
Colon:	diverticulitis
	carcinoma
	ulcerative colitis
	stercoral ulceration
	anastomotic leak
Small bowel:	strangulation, obstruction, vascular occlusion
	trauma
Gall bladder:	cholecystitis
Uterus:	traumatic perforation

Pathogenesis of peritonitis

1. Spillage of bowel flora into the peritoneal cavity is accompanied by substances such as mucus, bile, gastric juice, enzymes, blood and faecal compounds which cause inflammation and enhance the virulence of potential pathogens
   (injections of organisms into the peritoneal cavity of animals does not cause peritonitis unless compounds such as mucin or barium are added)
2. Aerobic and anaerobic bacteria act synergistically. Anaerobes, in particular *Bacteroides fragilis*, play the leading role in pus formation
3. Inflammatory peritoneal exudate causes adjacent bowel, omentum and mesentery to become glued together. Localisation of the infection may occur, particularly in one of the recesses of the peritoneal cavity, causing
   a. pelvic abscess
   b. subphrenic abscess
   c. paracolic abscess
4. Initial hypermotility of bowel is soon followed by paralysis (ileus). Fluid shifts cause a fall in circulating blood volume
5. Release of endotoxins, mostly from aerobic Gram-negative rods such as *Esch. coli*, causes septicaemia

Clinical features of peritonitis

1. Abdominal pain, guarding and rebound tenderness (peritonism)
2. Fever
3. Absent bowel sounds; later, abdominal distension and vomiting
4. Signs of septicaemia
5. If the infection localises, abdominal masses may be palpable

Diagnostic methods

1. Peripheral white cell count is usually elevated
2. Blood cultures may be positive, usually with anaerobes
3. Plain X-ray of the abdomen may reveal gas under the diaphragm and ileus
4. Peritoneal tap may be useful
5. Abdominal ultrasound and computerised tomography are useful for detecting localised infections

Treatment

1. Fluid and electrolytes
2. Surgery to correct the anatomical defect, remove foreign material and provide drainage
3. Antibiotics
   Creation of peritonitis in experimental animals has revealed that antibiotic therapy should be directed against:
   a. *aerobic bacteria*: to prevent sepsis
   b. *anaerobic bacteria*: to prevent abscess formation

Adequate regimens include

Penicillin ⎫
Aminoglycoside ⎬ used most frequently in the UK
Metronidazole ⎭
Clindamycin ⎫ used most frequently in the USA, may cause
Aminoglycoside ⎭ antibiotic-associated colitis
Newer cephalosporins: cefoxitin, latamoxef
Ticarcillin ⎫
Aminoglycoside ⎬ may cause bleeding
Chloramphenicol: effective but risk of bone marrow suppression or
agranulocytosis

Infection following peritoneal dialysis

— results from the entry of organisms into the peritoneal cavity via
the lumen or the skin tunnel of the catheter
— may recur frequently in chronic ambulatory peritoneal dialysis,
especially if aseptic procedures are not rigidly practised
— causes pain, tenderness, fever and turbid dialysate
— is caused by:
a. *Staph. epidermidis*
Strep. viridans
Diphtheroids
— cause mild peritonitis which usually resolves with addition
of antibiotics to rapidly cycled dialysate
b. *Staph. aureus*
— is less frequent, more severe. Systemic and topical
antibiotics are used
c. Gram-negative rods
— are less frequent, usually require removal of catheter
d. *Candida* species
— require removal of catheter

INFECTIONS OF THE BILIARY TREE

Cholecystitis

— is initiated by obstruction of the cystic duct by a stone, followed
by distention, tissue necrosis and bacterial proliferation.
Perforation may follow
— usually results in a mixed infection with *Esch. coli* predominating
— is treated with ampicillin, cephalosporins or combinations which
include aminoglycosides

Cholangitis

Follows obstruction of the common bile duct by stone, surgery,
tumour, pancreatitis or flukes.

Clinical, diagnostic and therapeutic features
1. Onset is usually abrupt with fever, liver tenderness and jaundice. Septicaemic shock may follow
2. Blood cultures are frequently positive with gram negative aerobic bacteria predominating
3. Peripheral white cell count is usually elevated
4. Serum alkaline phosphatase rises rapidly to high levels. Transaminases may be moderately elevated
5. Ultrasound will usually reveal stones and bile duct dilatation
6. Therapeutic levels of relevant antibiotics cannot be achieved in bile in the presence of jaundice but endotoxaemia can be ameliorated. Combinations including betalactams and aminoglycosides are most frequently used
7. Prompt operative decompression and exploration of the common bile duct is important
8. Temporary percutaneous biliary drainage may allow time for resuscitation in serious infection
9. If anatomical defects cannot be corrected, recurrence is frequent

LIVER ABSCESS

Causes of liver abscess include:
1. Amoebiasis
2. Ascending cholangitis
3. Portal bacteraemia/pyaemia
4. Systemic bacteraemia

In many patients, no source for the infection can be found but is presumed to have followed portal bacteraemia from a source of occult intestinal infection.
   Pyogenic liver abscesses usually contain multiple bacteria, particularly:
1. Anaerobes
2. *Streptococcus milleri*
3. Enteric Gram-negative rods and streptococci

Clinical and diagnostic features
1. Onset may be insidious with fever and right upper quadrant pain and tenderness
2. Jaundice is usually absent unless the abscess follows ascending cholangitis
3. Liver function tests reveal elevation of serum alkaline phosphatase and gammagluteryl transferase
4. Neutrophil count is usually elevated
5. Blood cultures often yield anaerobes or *Streptococcus milleri*
6. Chest X-ray reveals elevation of right hemidiaphragm and basal atelectasis

7. Isotopic liver scan, ultrasound or computerised tomography reveal liver defect
8. Percutaneous aspiration confirms the diagnosis and can be followed by catheterisation to establish drainage
9. Prognosis of pyogenic abscess is poor unless drainage is established by percutaneous catheter or surgery
10. Amoebic liver abscess usually responds to metronidazole

Some causes of liver scan defects include:

1. Liver abscess (amoebic or pyogenic)
2. Hydatids
3. Hepatoma
4. Metastases
5. Congenital malformation

Opportunistic infections

Infections, usually of low pathogenicity, occuring in hosts with immune defect.

Causes of immune defect
1. Congenital
2. Acquired
   a. Immunosuppressive therapy
      (i) Steroids
      (ii) Cytotoxics
      (iii) Irradiation
      (iv) Cyclosporin A
      (v) Antilymphocyte globulin
      (vi) Plasmapheresis
   b. Haematological malignancies, causing
      (i) neutropenia — acute leukaemia
      (ii) diminished cell-mediated immunity-lymphoma
      (iii) decreased antibodies — myeloma, chronic lymphatic leukaemia
   c. Splenectomy, asplenia
   d. Acquired immune deficiency syndrome (AIDS)
   e. Common variable immunodeficiency (late onset hypogammaglobulinaemia)
   f. Others: Diabetes, malignancy, alcoholism, uraemia, pregnancy, old age, burns, malnutrition
   The nature of the immune defect influences the type of opportunistic infection (Table 87).

Table 87

Nature of defect	Type of infection
Neutrophils	Gram-negative bacilli
	Gram-positive cocci
	Fungi
Immunoglobulins	Capsulated bacteria (pneumococcus,
Splenectomy	*H. influenzae*, meningococcus, streptococci)
Complement	*Staph. aureus*
	Enteroviruses
Cell-mediated immunity	Numerous classes of organisms (see below)

Neutropenia
— Risk of infection increases when neutrophil count is less than 1000/µl and is considerable with counts less than 200/µl
— Sites of infection include oral cavity, perianal region
— Bacteraemia is detected in 10–20% of febrile episodes, the source being frequently unknown but thought to be the result of damaged colonic mucosa
— Causes of bacteraemia include
Escherichia coli
Klebsiella/Enterobacter
Pseudomonas aeruginosa: has the highest mortality. Characteristic metastatic skin lesions (ecthyma gangrenosum)
Staphylococcus aureus: less common, lower mortality
Staphylococcus epidermidis ⎫ may contaminate venous lines
Corynebacterium JK ⎭
Bacillus cereus: unusual but causes lung cavitation, high mortality
Anaerobes are unusual causes of bacteraemia in the neutropenic host
Fungi (Candida, Aspergillus): not commonly isolated from blood cultures despite high incidence of organ involvement in prolonged neutropenia

Principles of management
1. Prophylactic oral antibiotics
   — reduce the incidence of febrile episodes by reducing colonic content of aerobic gram negative rods
   — may cause side effects, including nausea, overgrowth of resistant organisms
   — should be selected from antimicrobials which do not eliminate anaerobes (and therefore diminish colonisation resistance) of large bowel. Includes the following
     aminoglycosides: framycetin, neomycin, gentamicin
     colistin may cause nausea
     naladixic acid may have adverse effects
     cotrimoxazole also reduces the incidence of pneumocystis pneumonia in children
2. Prophylactic antifungal agents
   a. Mouth washes with amphotericin, nystatin
   b. Ketoconazole
3. Isolation
   a. Positive pressure ventilation of single rooms with clean air
   b. Gowns, gloves, masks, handwashing
   c. Washing in antiseptics, viz. hexachlorophane, chlorhexidine
   d. Avoidance of unheated food and drinks

4. Febrile episodes
   — should be treated promptly after collection of blood cultures
   using a combination of synergistic and preferably
   antipseudomanal antibiotics as follows:
   Betalactam: carbenicillin, ticarcillin, piperacillin, mezlocillin,
   cefotaxime or latamoxef
                         PLUS
   Aminoglycoside: gentamicin, tobramicin or amikacin
5. Persistant fever despite antibiotics poses a real problem in these
   patients. Methods of management include:
   a. *Systemic antifungals*: Ketoconazole, amphotericin B
   b. *Granulocyte transfusions*
   c. *Search for source of infection*: includes perianal lesions,
      venous lines
6. Pulmonary infiltrates
   — may be due to bacteria or fungi (especially *Aspergillus*)
   — cannot readily be investigated by bronchoscopy or
   percutaneous aspiration because of thrombocytopenia

Hypogammaglobulinaemia
— Increases susceptibility to certain groups of microorganisms, viz.
Haemophilus influenzae ⎫
Streptococci   ⎪ Recurrent sinopulmonary and pyogenic
Pneumococci   ⎬ infections with bronchiectasis as a
Staphylococcus aureus ⎭ frequent sequel.
Giardia lamblia: diarrhoea, malabsorption
Enteroviruses: chronic meningitis
   Treatment consists of replacement therapy with pooled adult
gamma globulin together with early treatment of infection

Splenic dysfunction
— Caused by splenectomy, sickle cell disease, congenital asplenia
— Splenic function includes:
   a. non-specific phagocytosis
   b. antibody production
      Spleen contains 25% of total lymphoid tissue
   c. production of tuftsin
      A polypeptide which coats neutrophils and enhances their
      ability to engulf bacteria
— Risk of infection is greatest in sickle cell disease and following
  splenectomy for thalassaemia and Hodgkin's disease
— Causes of infection include
  Capsulated bacteria, especially *Strep. pneumoniae*, fatal in
  50–70%
  Staph. aureus
  Protozoans: malaria, Babesia

— Prevention includes
  1. Delay splenectomy until 5 years of age
  2. Continous oral penicillin in young splenectomised children
  3. Pneumococcal vaccine

Defective cell-mediated immunity
Infections may be due to infectious agents from any class (Tables 88 and 89)

Table 88

Parasites	Fungi	Bacteria	Viruses
Pneumocystis	Cryptococcus	Listeria	Herpes simplex
Toxoplasma	Aspergillus	Legionella	Cytomegalovirus
Cryptosporidium	Candida	Mycobacterium	Varicella-zoster
Strongyloides		Nocardia	Hepatitis B
			Papovaviruses

Table 89 Organ involvement of infections associated with defects in cell-mediated immunity

Organ	Lesion	Organism
Lung (the organ most frequently involved)	Diffuse lesions causing hypoxia	Pneumocystis Cytomegalovirus
	Segmental lesions, occasionally cavitating	Mycobacterium Aspergillus Cryptococcus Nocardia Legionella Strongylioides
Brain	Meningitis	Cryptococcus Listeria
	Encephalopathy, space-occupying lesions	Toxoplasmosis Listeria Polyomavirus (PML) Lymphoma (due to EBV) Herpesviruses
Skin	Vesicular, crusting lesions	Herpes simplex Herpes zoster
	Proliferative lesions	Warts Kaposi's sarcoma
Mucous membranes	Ulcers	Herpes simplex Candidiasis
Gastrointestinal	Diarrhoea	Cryptosporidiosis Giardiasis
Eye	Retinopathy	Cytomegalovirus Candida Toxoplasma

The infections and their clinical manifestations are related to:
a. Duration of immunosuppression
b. Degree of immunosuppression
c. Presence of pre-existing latent infections (e.g. *Mycobacterium,
   Toxoplasma, Cytomegalovirus*)
The time course of infections in prolonged immunosuppression is
as follows:

First month: Mucocutaneous herpes simplex
                    Oral candidiasis
                    Pneumococcal pneumonia

1–4 months: Cytomegalovirus fever and pneumonia
                    Pneumocystis pneumonia

                    Aspergillosis
                    Nocardiosis
                    Listeriosis
                    Herpes zoster
                    Tuberculosis
                    Warts

After 4 months: Cryptococcal meningitis
                    Cytomegalovirus retinopathy
                    Progressive multifocal leucoencephalopathy
                    (PML)
                    Cryptosporidiosis

Fever, infections of travel

Body temperature is usually maintained between 35.8 and 37.2°C, being lowest in the early morning and highest in the late afternoon (diurnal variation).

A thermostatic mechanism in the anterior thalamus receives receptor information and influences efferent tracts to produce vasoconstriction and shivering or vasodilatation and sweating.

Mechanisms for fever production include
1. Pyrogens
2. Neurohypophyseal disorders
3. Diminished sweating (heat stroke)

Mechanism of fever production by pyrogens
1. Numerous compounds, usually of high molecular weight act as 'exogenous' pyrogens, including bacterial endotoxins, viruses, Gram-positive bacteria, antigen-antibody complexes, pyrogenic steroids
2. In response to stimulation with these compounds, circulating and fixed bone marrow-derived phagocytes (monocytes, macrophages) synthesise endogenous pyrogen, a protein of mol wt. 15 000, identical to interleukin-1, released over a period of hours
3. Endogenous pyrogen induces synthesis of prostaglandins, monoamines and possibly cyclic AMP in the anterior hypothalamus
4. From the anterior hypothalamus, information is transmitted to the posterior hypothalamus and the vasomotor centre
5. The ability of antipyretics such as aspirin to reduce fever is proportional to their ability to inhibit prostaglandin synthesis

The major causes of fever
1. Infection
2. Trauma, haematoma formation
3. Neoplasia
4. Hypersensitivity, drugs
5. Collagen-vascular diseases
6. Venous thrombosis, pulmonary embolism

Unusual causes of fever
1. Periodic diseases, especially familial Mediterranean fever
2. Granulomatous diseases of unknown aetiology
    a. Crohn's disease
    b. Sarcoidosis
    c. Granulomatous hepatitis
3. Whipple's disease
4. Self-induced (factitious) fever

Pyrexia of unknown origin
Defined by Beeson (1961) as an illness of at least 3 weeks' duration, with fever (> 38.3°C) and no established diagnosis after one week of hospital investigation.

Investigation of fever
1. *Cultures*: Blood, urine, sites of infection
2. *Haematology*: Full blood count, ESR, bone marrow biopsy
3. *Plasma biochemistry*, including tests of renal and hepatic function
4. *Serum antibodies* to appropriate bacterial and viral antigens and to certain tissue antigens (e.g. antinuclear antibodies, rheumatoid factor)
5. *Organ imaging*, including X-ray of the chest and other organs as indicated by clinical features and above investigations
6. *Biopsy* of tissues which reveal abnormalities

Causes of fever in hospitalised patients
1. Thromboembolism
2. Pneumonia
3. Renal tract infection
4. Drugs
5. Blood transfusion
6. Surgical wound infection

Causes of fever in returned travellers:
1. Malaria
2. Typhoid
3. Hepatitis B
4. Dengue
5. Amoebiasis
6. Viral haemorrhagic fevers

TYPHOID FEVER (enteric fever)
Typhos (Greek) = smoke, foul smelling. The term 'enteric fever' was introduced in 1869 when the aetiology was established and the disease differentiated from typhus.
   Salmonella typhi is an exclusively human pathogen which is usually transmitted by food contaminated by carriers of the organism

Pathogenesis
1. After ingestion, *S.typhi* invades bowel wall and multiplies in phagocytes of mesenteric nodes before causing bacteraemia
2. Foci of infection are produced in the liver, spleen, and lymphoid tissue of the intestine, occasionally in the kidneys, lung, bones and joints. Continous bacteraemia follows
3. Endothelial proliferation eventually causes vessel encroachment and necrosis. Peyer's patches ulcerate
4. *S.typhi* in liver enters bile and faeces

Clinical features of typhoid fever
1. Incubation period 10–14 days, insidious onset with chills, headache, cough, occasionally psychosis, meningism, pneumonia
2. Relative bradycardia in the presence of fevers of 38–39°C
3. Crops of transient erythematous macules (rose spots) may appear on the upper abdomen. Splenomegaly in 40–60%
4. Abdominal pain, distension. Diarrhoea, occasionally bloody
5. Prolonged course (untreated) which may be complicated by
   a. Toxaemia: hyperpyrexia, myocarditis, leukopenia
   b. Ulceration of intestinal lymphoid tissue, causing haemorrhage or intestinal perforation.
   c. Relapse after initial clinical response in 10–20% of patients, despite antibiotics
   d. Localised infections in bones, endocardium, lung
   e. Chronic carrier state, defined as faecal excretion of *S.typhi* persisting for more than one year. Occurs in 1–3% of patients, more commonly in old age and in females, due to associated biliary disease (cholelithiasis, schistosomiasis)

Diagnostic methods
1. Positive blood cultures
   More than 80% in first week, 50% in second and third weeks
2. *S.typhi* in faeces
   in 50% after the first week
3. *S.typhi* O and H agglutinins (Widal test)
   rise in titre in the second week but there are numerous problems with the test, including
   a. anamnestic response to antigens shared with other salmonellae
   b. antibodies due to prior immunisation
   c. blunted immune response following antibiotic therapy

Treatment
1. *Chloramphenicol*: 50 mg/kg/d for at least 2 weeks
2. *Alternatives*:
   a. amoxycillin 100 mg/kg/d
   b. sulphamethoxazole-trimethoprim 4–8 tablets per day

Table 90 Viral Haemorrhagic Fevers

Virus group	Disease	Mode of transmission	Mammalian reservoir	Distribution
Poxvirus	Smallpox	Air	Man	Eradicaticated
Flaviviruses	Yellow fever	Mosquitoes	Man / Monkeys	Tropical Africa and S. America
	Haemorrhagic, dengue		Man	All tropical countries
	Chikungunya fever		? Monkeys	India, S.E. Asia, Africa
Bunyaviruses	Rift Valley fever	Mosquitoes	? Rodents	Africa, causes disease in sheep and cattle.
Hantaviruses (members of the Bunyavirus group causing nephropathy)	Crimean H.F.	Ticks or mites	Small rodents	USSR, Bulgaria
	Kyasanur Forest disease		Small rodents	India
	Omsk H.F.		Muskrats	Siberia
	Korean H.F.		Field mice and voles	Korea, USSR, Japan
Arenaviruses	Argentinian H.F.	Contact with rodents or their excreta	*Calomys* spp	Argentina
	Bolivian H.F.		*Calomys callosus*	Bolivia
	Lassa fever		Multimammate rat	N.W. Africa
Filoviruses	Marburg fever	Unknown	Unknown	Africa
	Ebola fever			S. Sudan / N.E. Zaire

HAEMORRHAGIC FEVERS

Overwhelming infections causing bleeding tendency, especially into the skin:

1. Zoonotic RNA viruses
   — viral haemorrhagic fevers (see Table 90)
2. Rickettsiae
   — typhus, spotted fevers
3. *N.meningitidis*
   — Waterhouse–Friderichsen syndrome
5. Spirochaetes
   — leptospirosis, relapsing fevers

Hospital infections

Infections acquired in the hospital, also known as *nosocomial* infections (*nosocomion* (Greek) = a hospital), include the following:
1. Infections in a compromised host, usually due to normal body flora:
   a. Postoperative wound infections
   b. Bacteraemia along sites of venous access
   c. Renal tract infection following catheterisation
   d. Pneumonia following endotracheal intubation
   e. Infection following neutropenia and diminished cellular immunity (see chapter on Opportunistic Infections)
2. Transmissible diseases acquired from attendants, or other patients:
   a. via the respiratory route (smallpox, varicella-zoster, measles, rubella, common colds, influenza etc)
   b. via direct contact (herpes simplex, enteric pathogens etc)
3. Infections acquired from blood or its products

SURGICAL WOUND INFECTIONS

Pathways of surgical infection

Most infections come from the patient's normal flora rather than that of the attendants.

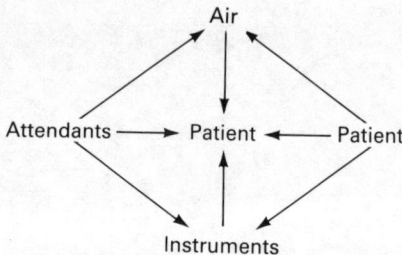

1. *Skin organisms*
a. *Staphylococcus aureus* causes infection in 1–5% of clean
   surgical wounds
b. *Staphylococcus epidermidis* causes infection in
   (i) the cement used in prosthetic orthopaedic surgery
   (ii) cardiac prostheses
   (iii) CSF shunts
c. *Streptococcus pyogenes* is a rare but serious cause of
   postoperative sepsis

2. *Mucosal organisms*
— cause infections following mucosal incision
— create mixed infections with pyogenic anaerobes playing a
   leading in pathogenesis, viz.
   Anaerobes: *Bacteroides fragilis*, other *Bacteroides* species,
                 anaerobic cocci
                 PLUS
   Aerobes: *Esch. coli*, streptococci, staphylococci

3. *Environmental organisms (water, disinfectants)*
Pseudomonas aeruginosa, occasionally other Gram-negatives.

Factors influencing the incidence of postoperative wound infection

1. *Virulence of organism*

2. *Degree of operative contamination*
a. Site of mucosal incision: Degree of contamination is in the
                               order colon > appendix > vagina
                               > stomach
b. Preoperative carriage of   The incidence rises during
   S.aureus: prolonged hospitalisation
c. Infections at other sites of the   e.g. boils, bedsores, varicose
   body: ulcers, wounds from previous
                               surgery
d. Preoperative shaving: causes folliculitis. Shaving should
                               be performed just before surgery
e. Prolonged surgery   All slightly increase the risk of
   Long incisions   operative contamination
   Crowded operating theatres

3. *Diminished host defences*
a. Poor haemostasis and   i.e. poor surgical technique
   tissue handling:
b. Presence of foreign   e.g. cement, Teflon, shunts
   bodies, prostheses:
c. Drains   promote the removal of blood and
                               exudate but also act as a portal of
                               entry for infection

d. Old age
e. Certain anatomical sites: e.g. axilla, groin, lower limb
f. Vascular disease
g. Diabetes
   Malignancy
   Malnutrition
   Steroid therapy

Prevention of surgical wound infection
1. Adequate sterilisation of instruments
2. Theatre design plus filtered air conditioning
3. Drapes, gowns, masks, gloves
4. Handwashing
5. Good surgical technique
6. Antiseptic skin preparation
7. Antibiotics given before surgery for
   a. breaching of mucosal surfaces
   b. introduction of foreign bodies
   c. vascular disease of lower limbs
8. Minimal movement and numbers of attendants

INFUSATE-RELATED INFECTIONS

1. Blood or blood products

a. Viruses
Hepatitis B: previously common, now prevented by screening donor blood for HBs Ag
Hepatitis non A non B: causes clinical hepatitis in 1% of patients receiving blood transfusions
Cytomegalovirus: is an occasional cause of fever after transfusion following large volumes of fresh blood
Epstein-Barr virus: may rarely follow transfusion
Lymphadenopathy-associated virus (LAV, HTLV III) may cause acquired immune deficiency syndrome especially in haemophiliacs
Human serum parvovirus/Human T-cell leukaemia virus (HTLV I): Transmission of these agents has recently been described

b. Protozoans
Malaria, Trypanosomiasis

c. Bacteria
Syphilis
   Prevention of infection by blood/blood products is diminished by:
1. Using only voluntary donors
2. Questioning potential donors about previous hepatitis, malaria, homosexuality
3. Screening blood for presence of hepatitis B surface antigen, antibodies to syphilis, cytomegalovirus (transplant recipients)

4. Avoiding pooled products (especially fibrinogen) whenever possible
5. Heat-treating albumin and fector VIII concentrates

2. Intravenous solutions
Solutions used intravenously may be contaminated in one of two ways:
a. by environmental Gram-negative rods (*Pseudomonas, Enterobacter*) during the course of manufacture (commercial or hospital)
b. by aerial and skin organisms (staphylococci, corynebacteria, fungi) during connection to giving sets or the introduction of additives
Sepsis due to contamination of manufactured solutions is prevented by:
a. Quality control following manufacture, viz.
   (i) membrane filtration of 1–10% of the batch followed by prolonged culture
   (ii) detection of endotoxins by rabbit test or Limulus lysate
b. Inspecting bottles before use
c. Introducing only one additive per bottle, then refrigerating before using as soon as possible
d. Changing i.v. sets every 24–48 hours

VASCULAR 'ACCESS' INFECTIONS

Bacteraemia frequently follows placement of the following devices in the circulation
Venous cannulae: central and peripheral, many purposes
Arterial cannulae, for measuring blood pressure
Transvenous pacemakers, balloon pumps, Swan-Ganz catheters, for monitoring and treating cardiac dysfunction
Arterio-venous shunts and fistulae, used mostly for haemodialysis in chronic renal failure
Umbilical catheters are used in neonates
   Organisms associated with access infections include
Staphylococcus epidermidis is the most frequent isolate, low pathogenicity unless it colonises cardiac and other implants
Staphylococcus aureus may cause septic thrombophlebitis, endocarditis, osteomyelitis, septic arthritis
Corynebacteria have low pathogenicity
Candida species occasionally cause retinal lesions
Gram-negative bacilli: less common but may cause septic shock
   Methods of prevention include the following
1. Use aseptic insertion techniques, record the date of insertion
2. Avoid cannulating veins at skin flexures
3. Inspect the insertion site daily, apply topical antiseptics
4. Remove the line as soon as possible

CATHETER-ASSOCIATED RENAL TRACT INFECTION

Bacteriuria complicates vesical catheterisation in 40–50% of patients after 7 days.

Consequences of infection
1. Pyelonephritis
   — may be associated with calculus formation if urea-splitting organisms are present
2. Bacteraemia
   — occurs in 1–2% of all catheterisations and is the commonest cause of Gram-negative septicaemia in hospital practice
3. Persistent bacilluria

Methods of prevention
1. Aseptic insertion techniques
2. Careful fixation of the catheter helps to prevent movement
3. Closed drainage techniques which keep the catheter and drainage bag in permanent continuity
4. Systemic antibiotic administration prevents infection for several days but resistant organisms eventually colonise
5. Bladder lavage with antibacterial agents is usually unsuccessful
6. Changing the catheter does not reduce the incidence of infection

LOWER RESPIRATORY TRACT INFECTION

Hospital-acquired infection of the lower airways has the following causes:
1. Contamination of nebuliser equipment
   — is usually due to environmental Gram-negative bacteria
   — is prevented by using sterile solutions and sterilising nebuliser equipment between patient use
2. Endotracheal intubation/tracheostomy
   — results in colonisation of the lower airways with Gram-negative bacteria in the majority of patients within a few days
   — causes pneumonia in a minority of patients

CONTAINMENT OF TRANSMISSIBLE INFECTIONS

Patients with transmissible diseases may transmit such infections to susceptible patients and staff if methods of containment are not used.
   The methods used depend upon the nature of the infection and its routes of transmission, viz.

1. Frequently fatal infections warranting high grade isolation
Smallpox
Plague
viral haemorrhagic fevers (see Table 90)
Anthrax

2. *More common infections transmitted by air*
Tuberculosis
Varicella-zoster
Diphtheria
Measles
Influenza
Pertussis

3. *Infections transmitted by faeces and fingers*
   Salmonella, shigella
   Enteroviruses

4. *Infections transmitted by blood and secretions*
   Hepatitis B
   Hepatitis non A non B
   AIDS

5. *Staphylococcal infections*

6. *'Coliform' infections*

Principles of infection containment
1. Patients should be nursed in single rooms if the infection is transmissible by air. Window should be fitted with an exhaust fan, door should be kept closed
2. Infectious patients nursed in common ward areas should be identified
3. All attendants should wash their hands after touching the patient
4. Susceptible and exposed staff should be immunised if vaccines are available, viz.
   Diptheria
   Pertussis
   Hepatitis B
   Tuberculosis
   Smallpox
5. When the patient leaves the ward, the bed and its environment should be cleaned ('terminal' cleaning)
6. Gowns and gloves should be worn when handling blood and exudate of patients with hepatitis B and non A, non B
7. Patients with enteric infections should use a toilet set aside for their use. If they cannot, bed pans should be put through an effectively operating pan sanitizer or the faeces covered with a phenolic disinfectant before sluicing
8. All specimens collected from patients with infections transmissible by blood should be 'flagged' before being sent to the laboratory

INFECTIONS TRANSMITTED TO PATIENTS BY STAFF

Staff may occasionally transmit infections to patients, as follows:
Enteric pathogens: Staff with diarrhoea should not work
Hepatitis B: Dentists and surgeons who are carriers may transit
hepatitis B
Rubella: Subclinical rubella in maternity staff may pose a threat to
pregnant women. Staff working in such areas should be tested
before employment
Herpes simplex: Staff with cold sores should not work in neonatal
units or other high risk areas
Respiratory viruses: are easily spread from staff to patients but
prevention is a problem. Staff with any evidence of respiratory tract
infection should be excluded from very high risk areas
Multiresistant
Staphylococcus aureus
is occasionally carried by staff working in areas where patient
colonisation rates are high. Skin disease (dermatitis, paronychia)
predisposes

Sterilisation and disinfection

Definitions
Sterilisation: Destruction or removal of all living microbes
Disinfection: Destruction of *pathogenic* organisms, not including bacterial spores, usually by chemicals (disinfectants)
Antisepsis: Prevention of infection of tissues or body surfaces by applied non-antibiotic chemicals (antiseptics)

Methods
1. Heat
   — Rapid, reliable
   — Destructive
   — Best method for small, heat-resitant objects
2. Chemicals
   — Slow, difficult to control efficacy
   — May be corrosive
   — Used for heat-sensitive objects, large surfaces, skin
3. Filtration
   — Requires membrane filtration apparatus
   — Can only be used for liquids
4. Irradiation
   a. ionising irradiation is reliable but expensive
   b. ultraviolet irradiation is still used in rooms despite doubts about its efficacy

HEAT

1. Pasteurisation
Destruction of pathogens in heat-susceptible fluids (especially milk) by mild heat treatment, e.g. 63°C for 30 minutes, 78°C for 15 seconds. Does not sterilize.

2. Tyndallisation
Repeated steaming (at 100°C) of culture media on each of three successive days, allowing spores to germinate and to be susequently killed.

3. Boiling or steaming (at 100°C)
Kills vegetative organisms rapidly. Spores survive.

4. Autoclaving
Exposure of heat-resistant materials to saturated steam under increased pressure and therefore a temperature exceeding 100°C. Steam — provides seven times the available heat of boiling water (at 100°C)
  — penetrates well
  — kills by coagulating protein

Table 91 Relationship of steam pressure to temperature

| | Pressure of steam | | Temperature achieved (°C) | Time required to kill spores (minutes) |
kPa	lb/in³	mmHg		
101	15	760	121	15
131	20	1000	126	10
202	30	1520	134	2

The process of autoclaving
a. The objects to be sterilised are wrapped in protective materials which are pervious to steam
b. The pressure vessel (autoclave) is loosely loaded with the wrapped objects and sealed
c. Air is removed by
   — downward displacement as the steam is introduced, OR
   — the application of a vacuum before the steam is introduced
d. The sterilisation time is calculated to allow adequate penetration and killing of the load in question
e. The temperature and time of each heat cycle is recorded
f. The load is cooled and dried
g. Tapes which change colour after adequate time and temperature (Bowie–Dick test tape) are placed on each packaged item, verifying by the colour change that adequate conditions have been supplied
h. Periodic checks of autoclave efficacy are made using
   (i) test spores (*Bacillus stearothermophilus*)
   (ii) thermocouples placed in typical loads

5. Dry heat
Dry heat penetrates slowly and micro-organisms are considerably more resistant to it than moist heat
   Much higher temperatures and longer exposure times are required to sterilize (e.g. 160°C for 45 minutes)
   Two methods are used

a. Incineration
— used for disposal of dressings, laboratory media and human tissues using oil-fired hospital incinerators but may cause unacceptable environmental pollution
— is much used in microbiology laboratories (wire loops, glassware)

b. Hot air ovens
— should be thermostatically controlled and contain a fan to circulate the air
— used for objects which cannot tolerate moisture
— cannot be used for bulky items (heat penetrates poorly) or objects damaged by high temperatures (rubber, plastic)
— mostly used for glassware, oils, powders

CHEMICALS

Destruction (or the prevention of growth) of organisms by chemicals is used in the following circumstances:
1. *Environment*: Disinfection of excreta, floors, furniture, linen and fabrics
2. *Instruments* / *Equipment* } Sterilisation of heat-sensitive objects in contact with patients
3. *Skin and wounds*: Removal of pathogens (antisepsis)
4. *Food* / *Medications* / *Solutions* } Preservation (i.e. prevention of spoilage by the multiplication of environmental organisms)
5. *Water supplies*: Removal of pathogens
6. *Vaccines*: Destruction of organisms or denaturation of toxins

Classes of chemicals

1. Halogens
— are capable of killing bacteria, spores and viruses
— are only moderately toxic
— are very readily neutralised by organic matter
— include
Chlorine, hypochlorites, chloramines: used in water supplies, swimming pools, wounds, *clean* utensils
Iodine: used in alcohol as a tincture, for skin antisepsis
Iodophors: organic iodine, povidone iodine. Used as an antiseptic and handwash

2. *Alcohols*
— kill vegetative bacteria rapidly but have no action on spores
— are mostly used for skin disinfection, although there is a risk of burns if diathermy follows their use
— includes
  a. ethyl alcohol, used at 75%, in water (which is necessary for bactericidal activity)
  b. isopropylalcohol; more effective than ethanol

3. *Metallic salts*
— are slowly-acting but bacteristatic even in very low concentrations
— used as preservatives, no longer used clinically

4. *Alkylating agents*
— are capable of killing bacteria, spores and viruses (except slow viruses) and are the only group of compounds which are an acceptable chemical alternative to usual heat treatment
— include
  a. formaldehyde
    — Prepared as formalin, a 38% aqueous solution of formaldehyde gas
    — Used for instruments and machines such as haemodialysers and pump oxygenators but irritating, causes hypersensitivity
  b. gluteraldehyde
    — Prepared as an alkaline aqueous solution, less irritating than formalin, used for instrument sterilisation
  c. ethylene oxide
    — An inflammable gas, used as a 10% mixture in carbon dioxide at about 50°C in a suitable chamber. Humidity must be adequate
    — Penetrates rubber and plastic, used for anaesthetic apparatus and catheters
  d. propiolactone
    — A liquid which rapidly hydrolyses to hydracrylic acid in aqueous solutions. The vapour is used for gaseous sterilisation

5. *Phenols*
— include tar acids and semi-synthetic compounds in soap solutions
— are somewhat toxic with an unpleasant smell
— are used mostly as 'general purpose' disinfectants of contaminated inert surfaces
— have no activity against spores but are little inactivated by protein and organic matter

6. *Hexachlorophane*
— is a synthetic diphenyl derivative of phenol
— is active only against gram positive cocci
— has had much use as a handwash and in the prevention of staphylococcal infection in nurseries
— is potentially neurotoxic if applied to broken skin

7. *Quaternary ammonium compounds ('QATS')*
— are active against Gram-positive cocci but only feebly against Gram-negative rods and ineffective for spores and mycobacteria
— are effective, cheap and non-corrosive cleaning compounds
— are readily contaminated with environmental Gram-negative rods, particularly *Pseudomonas* species

8. *Dyes*
— include anilines and acridines which are active against Gram-positive cocci but which cause staining

9. *Chlorhexidine*
— is a synthetic compound of low toxicity, active against Gram-positive and somewhat active against Gram-negative bacteria but not spores, mycobacteria or viruses
— is inactivated by soap but not by the quaternary ammonium compound, cetrimide, with which it is often combined
— is used as a skin disinfectant and handwash

10. *Oxidising agents*
— include hydrogen peroxide and potassium permanganate
— are very readily inactivated by organic matter and have weak antibacterial activity

Principles of disinfectant use
1. Only objects which cannot be heat-treated should be disinfected by chemicals
2. It is important to remove grease and protein before use
3. Many substances, including hard water, soaps, detergents, chemicals, and plastics may inactivate disinfectants. Follow the manufacturer's instructions before use
4. Dilution of concentrates should be accurately measured, marked with an expiry date and used as soon as possible
5. Dilute solutions of some chemicals (particularly quaternary ammonium compounds) support the growth of environmental Gram-negative rods. For use on body surfaces, such dilutions should be made aseptically
6. Disinfectants should be appropriately chosen for their purpose, having regard for cost, antimicrobial range, toxicity and corrosiveness
7. Use of disinfectants for general cleaning is a waste of money

Disinfection/sterilisation in special situations

1. Heat sensitive instruments
a. Gluteraldehyde
b. Ethylene oxide
c. Formaldehyde
d. Steam at low pressure, with or without formaldehyde
e. Irradiation

2. Slow viruses
These viruses resist boiling, alkylating agents, alcohol and
ultraviolet irradiation.
a. Autoclaving for one hour at 121°C
b. Hypochlorite (5%)
c. Permanganate (0.03%)
d. Phenolics
e. Iodine solutions

3. Vaccines
a. Formalin
b. Heat
c. 8 molar urea
d. Propiolactone

Immunisation

The process of inducing immunity artificially. The term 'vaccination', which originally referred to the inoculation of vaccinia virus to render individuals immune to smallpox, is now often used for immunisation generally.

Immunisation may be active or passive

1. Active
Production of antibody in response to the administration of antigens which may be:

Live attenuated organisms
— produce long lasting immunity
— include many viral vaccines

Killed whole organisms
— produce immunity which may be incomplete

Fractions of organisms
— cause less frequent reactions

Toxoids
— are produced from bacterial exotoxins by treatment with formaldehyde
— are usually given with an adjuvant

2. Passive
Provision of temporary immunity by the administration of preformed antibodies

a. Pooled adult globulin
— is 15–18% protein globulin obtained by cold ethanol fractionation of large pools of adult human plasma
— contains sufficient antibody to protect subjects from:
  (i) hepatitis A
  (ii) measles

b. Specific immune serum glublin
— is immunoglobulin obtained from human plasma screened for high titres of antibodies to certain viruses, viz.

  (i) hepatitis B
(iii) varicella-zoster
(iv) rabies
 (v) vaccinia
— is used to give transient protection to unimmunised exposed
   subjects

c. *Antitoxins*
— are immunoglobulins obtained from antigenically stimulated
   animals or humans and used in the prevention or treatment of
   infections associated with toxin-producing bacteria, viz.
  (i) diphtheria
 (ii) tetanus (human antitoxin is preferred)
(iii) botulism
(iv) gas gangrene
— may cause serum sickness if the immunoglobulin has been
   obtained from animals

BACTERIAL VACCINES

1. Whole killed cells

Pertussis
— is included in the triple vaccine given initially at 2–3 months of
   age
— gives about 80% protection
— very rarely causes encephalopathy within 24 hours of the first or
   subsequent injections

Typhoid
— gives about 70% protection for an indeterminate period of time
— causes variable amount of pain at the injection site and fever
   some hours after the injection

Cholera
— gives limited protection for less than six months

Typhus and plague
— relatively weak immunogens but useful for high-risk groups

2. Surface components
Vaccines made from the surface components of bacteria are being
developed but are not yet in common use.

a. *Pneumococcus*
— a combination of 14 common types of capsular polysaccharide.
— poor immune response before the age of two.
— greatest use in certain patients with impaired splenic function,
   e.g., sickle cell disease, Hodgkins disease staged by
   splenectomy.

b. *Meningococcus*
— available for capsular types a and c, the usual cause of epidemics in Third World countries.
— the capsular antigen of type b, the usual cause of sporadic disease, is too poorly antigenic to use as a vaccine.

c. *Haemophilus influenzae*, type b
The capsular polysaccharide (polyribophosphate) fails to produce antibodies in children under about 18 months of age unless an adjuvant such as *Bordetella pertussis* is present.
   Outer membrane proteins (OMP's) of the cell wall are protective in experimental animals and may make useful vaccines.

d. *Bordetella pertussis*
Vaccines made from surface components are currently being evaluated.

e. *Pseudomonas aeruginosa*
A vaccine made from the surface components of the 16 serotypes has reduced the mortality from septicaemia in patients with burns.

f. *Escherichia coli*
Vaccines made from rough mutants with cell walls containing only core glycolipid can protect experimental animals from septicaemia.

3. **Live bacteria**

a. *BCG (Bacille de Calmette–Guérin)*
— is a strain of *M. bovis* rendered avirulent as a result of several years' subculture of the organism on glycerol bile potato medium
— injected intradermally causes only a small lesion which heals spontaneously, producing tuberculin hypersensitivity in most subjects
— produces some protective effect against pulmonary tuberculosis, the magnitude of which is still disputed
— when administered in early childhood reduces the incidence of miliary and meningeal disease

b. *Typhoid*
A strain of *S. typhi* which lacks the enzyme UDP-galactose-4-epimerase when swallowed undergoes 4 or 5 cell divisions and penetrates the cell wall of the small intestine before ceasing growth.
   Recent controlled trials reveal efficacy of greater than 90%.

4. **Toxoids**
Inactivated bacterial exotoxins combined with an adjuvant and given repeatedly provide longlasting immunity from diseases which are primarily a manifestation of toxin production.

a. *Corynebacterium diphtheriae*
— is a component of triple vaccine given at about 3 months of age
   as passively acquired maternal antibodies wane

b. *Clostridium tetani*
— is a component of triple vaccine
— after the initial course should be repeated about every 10 years or
   after sustaining a tetanus-prone wound

c. *Clostridium welchii*, type C (causing 'pig bel')

d. *Bacillus anthracis*

VIRUSES (in order of introduction to clinical medicine)

1. Smallpox
In 1796 Jenner demonstrated that cowpox inoculation protected a
boy from smallpox.
  The disease has now been eradicated, eliminating the need for
vaccination of civilian populations.
  The vaccine is made by harvesting vaccinia virus from the
exudative lesions produced by the inoculation of the skin of calves.
Bacterial contamination of the lymph is controlled by phenol and
glycerol.
  The vaccine is given intradermally, producing a vesicular lesion
and local lymphadenopathy. Protection lasts for 3 years.

2. Rabies
Pasteur demonstrated the efficacy of vaccination in 1884.
  Three types of vaccine have evolved for post-exposure
immunisation:
Semple vaccine
— 10% suspension of brain of sheep, rabbit or goat infected with
   passaged ('fixed') strain of rabies virus inactivated by phenol
— Given in daily doses for 14 days, then days 21 and 28
— Allergic and neurological reactions are frequent
Duck embryo vaccine
— Passaged strain propogated in duck embryo and inactivated by
   beta-propionolactone
— Less allergic encephalomyelitis
Tissue culture vaccine
— Rabies virus grown in human diploid tissue culture, inactivated
   by beta-propiolactone
— Much improved antibody response. Occasional hypersensitivity
— Should replace previous vaccines

3. Yellow fever
The live attenuated 17D strain of yellow fever virus is highly
effective, producing immunity that lasts at least 10 years. Side
effects are rare.

4. Influenza
Vaccines have been available but used irregularly for twenty years. Efficacy is hampered by the antigenic instability of the virus such that virus strains included in the vaccine must be reviewed annually.
Three types are in use:
Live attenuated: administered intranasally, have been used mainly in the USSR
Whole virus inactivated: made by growing current strains in chickembryo allantoic fluid, harvesting and inactivating with formalin
Purified surface (split) antigen: made by splitting the virus with ether. Cause less frequent reactions, especially in children
Priority for vaccination should be given to two groups
a. otherwise healthy individuals over the age of 65
b. adults and children with cardiac, renal and chronic lung disease, diabetes mellitus, anaemia and immunosuppression

5. Poliomyelitis
Two types of vaccine are available
Salk (inactivated): a formalin-inactivated mixture of the three serotypes of poliovirus
Sabin (live): attenuated strains of the three serotypes grown in tissue culture and given by mouth

6. Measles
Evolution of vaccines has been as follows:
a. A killed vaccine was originally introduced but when immunity waned, vaccinees exposed to natural measles developed severe and atypical disease
b. The live attenuated Edmonston B strain produced lasting immunity but caused high fever and rash
c. The Schwarz strain, further attenuated by 77 additional passages in chick embryo fibroblasts at 32°C provides lasting immunity with few side effects

Because measles is a disease only of humans, produces no carrier or latent state and can be prevented by vaccination, it could, like smallpox, be eradicated.

7. Rubella
A live attenuated vaccine (Cenderhill strain) was first introduced in 1969. A recently introduced strain, RA27/3, causes less side effects.

8. Mumps
A live attenuated vaccine (Jeryl Lynn strain) introduced in 1967 gives a high protection rate and few side effects.

9. Hepatitis B
Vaccines have been developed using inactivated, alum-adsorbed surface antigen derived from the blood of carriers. Three doses spaced 1 and 6 months apart produce good anti-HBs levels.

10. Varicella
Attenuated vaccines have been developed and are undergoing trials.

ADVERSE EFFECTS OF VACCINES

1. Minor toxicity
a. pain and swelling at injection site
b. fever, myalgia, headache
c. local skin ulceration and regional lymphadenopathy (especially with BCG and smallpox vaccination)

2. Faulty production resulting in
a. bacterial or viral (hepatitis B) contamination
b. undenatured toxin
c. presence of incorrect (and virulent) live agent

3. Allergy
a. to compounds used in vaccine manufacture, such as
 — egg (yellow fever, influenza)
 — antibiotics (neomycin)
 — mercurials
b. to the microbial antigen, resulting in
 — Arthus phenomenon at injection site
 — serum sickness
 — anaphylaxis

4. Neurological disease
It is generally assumed that these are hypersensitivity disorders because they follow vaccine use by about 10 days.
a. Brachial neuritis
 — usually involves fifth cervical nerve root with pain, paralysis and atrophy corresponding to the distribution of the nerve
 — occasionally involves entire brachial plexus
b. Landry–Guillain–Barré syndrome
c. Encephalomyelitis
 — is most frequently seen after rabies and smallpox vaccination
 — has a high mortality (30–50%)

5. Systemic disease following use of live agents
Although live vaccines contain organisms which should be sufficiently attenuated not to cause disease, they may do so in certain circumstances, viz.

a. invasion of normal host by certain vaccine viruses
Rubella causes arthralgia, especially in women
Poliovirus type 3 causes paralysis in less than one per million vaccinated
Measles has rarely been shown to cause subacute sclerosing panencephalitis
b. invasion in the presence of immunodeficiency
Vaccinia may cause progressive local ulceration if cell-mediated immunity is defective
BCG may cause disseminated disease if cell-mediated immunity is defective
Poliovirus: vaccine occasionally causes poliomyeltis in children with agammaglobulinaemia
c. transplacental infection
Vaccinia rarely causes generalised fetal vaccinia
Rubella has been detected in placenta but has not been demonstrated to cause malformation

6. Provocation of disease by vaccines
a. intramuscular injections provoke paralysis in that muscle group in subjects with prodromal poliomyelitis
b. killed typhoid vaccine may worsen the disease of subjects in the incubation period of typhoid fever

VACCINE MANUFACTURE AND FORMULATION

1. Live vaccines are prepared as lyophilised cultures to which must be added a suspending fluid. They must be stored at appropriate temperature to maintain potency
2. Aluminium compounds are used in some vaccines to enhance the immune response (ie, to act as adjuvants)
3. Multidose vials are useful for mass imunisation programmes. They may contain preservatives but should not be stored once opened
4. Vaccines containing killed pathogens must be carefully tested both in vitro and in animals to be certain that viable organisms do not remain
5. Wherever possible, vaccines must be tested for the efficacy (i.e. immunising capacity) of the product

PRINCIPLES OF VACCINE USE

1. Vaccines containing adjuvants must be given by deep intramuscular injection (anterolateral aspect of upper thigh, upper outer quadrant of buttock, deltoid muscle of upper arm)
2. After inserting the needle, the plunger of the syringe should be drawn back. If blood appears in the syringe, the needle should be re-inserted

3. Giving doses of a vaccine at less than the recommended interval may lessen the antibody response. On the other hand, intervals longer than those recommended do not lessen antibody response
4. Most antigens, including live virus vaccines, can be given simultaneously without blunting the immune response to any one of them or increasing the rate of adverse effects
5. Passively acquired antibody interferes with the replication of live viruses, therefore
   a. administration of live virus vaccines to infants before maternal antibodies have disappeared may yield poor immunity (this is a particular problem with measles vaccine given in the first year of life)
   b. live virus vaccines should not be adminstered for at least 6 weeks after administration of immune globulin
6. Live virus vaccines should not be given to
   a. pregnant women
   b. patients with immunodeficiency
7. Immunisation rates of less than 70% change the epidemiology of viral illnesses of childhood, rendering adult (and potentially more severe infection) more frequent. Government-sponsored programmes should aim for a high rate of immunisation
8. Decisions concerning immunisation are based on the expected benefit and the known risks of the vaccines

USE OF VACCINES IN VARIOUS GROUPS AT RISK OF INFECTION

1. Early childhood (see Table 92)

Diphtheria ⎱
Tetanus   ⎰ are combined in triple vaccine (DPT)
Pertussis

Poliomyelitis
Measles
Mumps

Rubella may be given to all children in early childhood or prepubertal girls

2. Travellers
The vaccines used depend upon the destination, the duration of stay and the life style of the traveller.

Typhoid
Cholera
Hepatitis A (pooled adult gamma globulin)

Hepatitis B

Tuberculosis

Table 92 Schedule of immunisation in childhood

Age	Disease	Agent
2 months	Pertussis, diphtheria, tetanus Poliomyelitis	Triple antgen Sabin oral
4 months	Pertussis, diphtheria, tetanus Poliomyelitis	Triple antigen Sabin oral
6 months	Pertussis, diphtheria, tetanus Poliomyelitis	Triple antigen Sabin oral
15 months	Measles, mumps	Combined live vaccine
18 months	Diphtheria, tetanus	Diphtheria and Tetanus, Toxoid (CDT)
At school entry	Diphtheria, tetanus Poliomyelitis	Diphtheria and Tetanus Toxoid (CDT) Sabin oral
12–14 years	Rubella	Live rubella vaccine

Pertussis vaccine should not be given to children:
1. with a previous history of neurological disease
2. with a family history of neurological disease
3. with a severe reaction to previous triple antigen (e.g. persistent screaming, vomiting, collapse, convulsions or fever > 39.5°C
4. over the age of 18 months (reactions are more frequent)

Yellow fever — equatorial Africa and South America

Rabies, following animal bites in endemic areas

3. Laboratory and hospital staff

Hepatitis B
Diphtheria
Tuberculosis
Influenza

Rubella, for non-immune staff working with pregnant women

Smallpox
Plague   for those working with these agents
Typhus

4. Patients at special risk

Influenza   cardiac, renal and chronic lung disease

Pneumococcus   following splenectomy

Tetanus   following tetanus-prone wound

Index